Springer Theses

Recognizing Outstanding Ph.D. Research

Aims and Scope

The series "Springer Theses" brings together a selection of the very best Ph.D. theses from around the world and across the physical sciences. Nominated and endorsed by two recognized specialists, each published volume has been selected for its scientific excellence and the high impact of its contents for the pertinent field of research. For greater accessibility to non-specialists, the published versions include an extended introduction, as well as a foreword by the student's supervisor explaining the special relevance of the work for the field. As a whole, the series will provide a valuable resource both for newcomers to the research fields described, and for other scientists seeking detailed background information on special questions. Finally, it provides an accredited documentation of the valuable contributions made by today's younger generation of scientists.

Theses are accepted into the series by invited nomination only and must fulfill all of the following criteria

- They must be written in good English.
- The topic should fall within the confines of Chemistry, Physics, Earth Sciences, Engineering and related interdisciplinary fields such as Materials, Nanoscience, Chemical Engineering, Complex Systems and Biophysics.
- The work reported in the thesis must represent a significant scientific advance.
- If the thesis includes previously published material, permission to reproduce this must be gained from the respective copyright holder.
- They must have been examined and passed during the 12 months prior to nomination.
- Each thesis should include a foreword by the supervisor outlining the significance of its content.
- The theses should have a clearly defined structure including an introduction accessible to scientists not expert in that particular field.

More information about this series at http://www.springer.com/series/8790

Jing Wang

CMOS-Compatible Key Engineering Devices for High-Speed Silicon-Based Optical Interconnections

Doctoral Thesis accepted by
the University of Chinese Academy of Sciences,
Beijing, China

Springer

Author
Dr. Jing Wang
Optixpan Semiconductor Inc.
Shenzhen, Guangdong, China

Supervisor
Prof. Fuwan Gan
Shanghai Institute of Microsystem
and Information Technology
Chinese Academy of Sciences
Shanghai, China

ISSN 2190-5053 ISSN 2190-5061 (electronic)
Springer Theses
ISBN 978-981-13-3377-4 ISBN 978-981-13-3378-1 (eBook)
https://doi.org/10.1007/978-981-13-3378-1

Library of Congress Control Number: 2018961718

This Springer imprint is published by the registered company Springer Nature Singapore Pte Ltd.
The registered company address is: 152 Beach Road, #21-01/04 Gateway East, Singapore 189721, Singapore

Supervisor's Foreword

Silicon photonics has attracted much attention in the past decades. Specially, in the recent five years, silicon photonics has moved from the research laboratory to manufacturing companies. This thesis written by Dr. Jing Wang was devoted to achieving a high-speed and large-scale silicon optical module for better optical interconnections and communications in the data center and metro-distance.

This thesis starts with Chap. 1, where the basis of silicon photonics is introduced. In Chap. 1, the motivation of silicon photonics for optical interconnections is discussed. Then, some progress on various silicon photonic components, devices, and modules is reviewed. Chapter 2 introduces the core of the high-speed silicon photonic module, i.e., silicon optical modulator. The design, simulation, fabrication, and characterization on two kinds of optical modulators are presented. Chapter 3 is about how to increase the interconnection capacity by using some optical multiplexing technologies. Some advanced integrated multiplexers on wavelength, polarization, and mode are developed. Chapter 4 addresses the problem of polarization controlling on a silicon photonic chip. Some CMOS-compatible polarization splitter-rotator are designed and fabricated, which can be used for coherent optical modules. Chapter 5 deals with light coupling from optical fiber to waveguide. Some simple and efficient edge-coupling solutions are discussed.

I hope that the efforts of Dr. Jing Wang will be useful to both specialists and newcomers. I also feel honored that his dissertation is selected in Springer Theses, proving our previous work on silicon photonics is meaningful and recognized. I have witnessed the evolution of this field from a privileged point of view and could accompany it from its infancy to full maturity. I look forward to further excitement ahead and new developments in silicon photonics.

Shanghai, China
September 2018

Prof. Fuwan Gan

Acknowledgements

First of all, I want to thank my three supervisors. They are Prof. Shichang Zou, Prof. Fuwan Gan, and Prof. Minghao Qi. I thank Prof. Zou for his teaching and caring in my study and life in the past five years. His rigorous, realistic, and meticulous character made me benefit a lot. He was conscientious and dedicated to the development of the semiconductor industry in China. I thank Prof. Gan for taking me into a fascinating discipline, silicon photonics, which makes me deeply addicted to it. Professor Gan is always so enthusiastic and vigorous. He always faces the pressure of work with confidence and optimism. Professor Gan gave me a lot of guidance and help in the design of active devices, and also gave me a lot of inspiration in managing the project. I also would like to thank Prof. Gan for sending me to the USA as a visiting student for one year, so that I can think deeply about my future. I also want to thank Prof. Qi, who was my supervisor in the USA. Professor Qi is a well-known expert in the field of optoelectronics. He is humorous, conscientious, and rigorous. He not only gives me a lot of guidance in the research work, but also cares about my life and makes me happy and substantial in the year studying in the USA.

I also would like to thank Dr. Zhen Sheng, who is very strict and rigorous and has rich and systematic knowledge on optics, for his guidance and modification in the research of arrayed waveguide gratings. At the same time, I would also like to thank Dr. Aimin Wu and Dr. Wei Li for their help and encouragement in my daily life.

I thank Dr. Hao Li, Dr. Zhiqi Wang, and Dr. Chao Qiu for their help in daily research. I also want to thank Wei Lin, Haiyang Huang, You Li, Ming Li, and Xin Chen for working together, talking about the future, and thinking about the future.

I thank Albert Peng and Le Li in Grace Semiconductor Manufacturing Co., Ltd., for their support on device processing.

I thank Dr. Ben Niu for his concern and help in my exchanging life in the USA. I thank Dr. Jian Wang and Dr. Juan Zeng for taking me to the supermarket to buy food and help me adjust to the life in the USA. I thank Dr. Yi Xuan for his support on device processing. I thank Justin Wirth for helping me to test the polarization controller. Also, I thank Kyunghun Han, Min Teng, Sangsik Kim, Feng Zhang, Xiaoxiao Xue, and others for their help in my work.

I thank Prof. Xi Wang, Prof. Yuehui Yu, Prof. Miao Zhang, Prof. Zhengxuan Zhang, Prof. Zengfeng Di, Prof. Xinhong Chen, Dr. Wenjie Yu, Dr. Xing Wei, Dr. Zhongying Xue, Dr. Zhongjian Wang, Dr. Dawei Bi, Dr. Bingxu Ning, Dr. Zhiyuan Hu, Mrs. Yuqiong Tang, Mrs. Lin Fan, Mrs. Tong Zhang, Mr. Dong Jing, and Mr. Min Zhou for their support and assistance in our research work.

I thank Mrs. Qi Luo, Mrs. Jia Xie, Mrs. Xiangbo Yu, and Mr. Tianhao Shao for their help in work, study, and life, and for training and supporting me when I worked as the chairman of the student union. This experience will benefit me in all my life. I thank Mrs. Weizhu Zhang for her strong support in literature search and help in my job hunting.

I thank many friends for their unselfish help to overcome many difficulties in learning and scientific research.

I also thank my parents for bringing me up. They have always been my utmost psychologic crutch. I will cherish my heart and remember in my life.

Finally, I want to thank my girlfriend, Dr. Xi Fan, who accompanied me and gave me a lot of comfort when I needed help.

I would like to dedicate this dissertation to my most beloved grandmother!

Contents

Chapter 1
Introduction

1.1 The Challenges of Moore's Law and the Bottleneck of Electrical Interconnection

With the rapid development of modern information industry and the advent of the Internet of things (IoT), cloud computing and big data era, the demand for information data capacity and processing speed is increasing rapidly. In the traditional microelectronics field, the integration and performance of the circuit are increasing in accordance with the "Moore's law" [1]. At the beginning of 2015, Intel released the Xeon D-1500 processor based on the Braswell architecture made by the latest 14 nm process. If in accordance with the original Tick-Tock development model, Intel should launch a new 10 nm process by the end of 2016. But the CEO in Intel has also recently talked about the 10 nm process and he just said they were still working on it and had not yet established a timetable. This is likely to mean that the opportunity to move to 10 nm at the end of 2016 is impossible. It can be seen that Moore's law is getting more and more serious and is unable to continue to develop rapidly as Moore predicted. Why does it happen? On one hand, the feature size is limited by the semiconductor processing technology and it is not possible to reduce unrestrictedly. When the size of the device is about 10 nm, there will be an obvious quantum effect, which will cause the failure of the transistor function. On the other hand, as the size of the device becomes smaller and smaller, the delay and loss of electrical interconnection are constantly highlighted [2]. Although copper has been used instead of aluminum and the low-K dielectric material has improved these performances to some extent, with the further shrinking of the device size, the problem of delay and loss can not be avoided. Moreover, with the increase of interconnection density and number of interconnection layer, the inter-layer crosstalk, electromagnetic interference, bandwidth limitation, excessive power consumption and heat dissipation bring new challenges to the microelectronics industry [3–6].

In the data center field, the structure of the processor becomes more and more complex when the performance of single-core processor is close to the physical lim-

J. Wang, *CMOS-Compatible Key Engineering Devices for High-Speed Silicon-Based Optical Interconnections*, Springer Theses,
https://doi.org/10.1007/978-981-13-3378-1_1

itation. At this point, the development of microprocessors has entered the multi-core era, and single-core processor has become the focus of research [7]. In addition, as the processor system frequency and size increases, the electrical interconnection technology based on electronic transport carrier and metal wires makes the information processing speed unable to meet the overall requirements of the system due to signal attenuation, reflection, signal crosstalk, electromagnetic interference, poor scalability, limited bandwidth, longer delay longer and large power consumption. Therefore, the electrical interconnection or switching has become a huge bottleneck for the development of multi-core processor, which seriously restricts the improvement of system performance [3–6].

1.2 Proposal of Silicon Photonics

Optical fiber communication has become a mainstream technology in long-haul high-speed communication. For the short-distance interconnection, optical interconnection technology and optoelectronic integrated circuits can be used instead of electronics and the photon works as the carrier of energy, which can achieve large capacity, high speed, little heat dissipation and no interference of information transmission. So it is considered to be an important development direction of short-distance and the inter-chip interconnection [8, 9]. Silicon photonics is one of the most important development directions of optical interconnections [10, 11].

Silicon photonics can use a standard CMOS technology to develop silicon photonic devices, and integrates silicon photonic devices and electronic circuits on the same silicon chip, where "Photon" is the main signal carrier. Why does silicon-based optical interconnection cause so much attention? Silicon has the advantages of low cost and high quality compared to other semiconductor materials. It has high thermal conductivity (about 10 times more than GaAs), high optical damage threshold (about 10 times more than GaAs) and high third-order optical nonlinear effects (e.g., Kerr effect and Raman effect are 100 times and 1000 times higher than standard fiber, respectively) [10]. In addition, silicon-on-insulator (SOI) material has a high refractive index difference, which is beneficial to the realization of ultra-small functional devices and improve the integration of chip [12]. Moreover, it can combine with mature silicon processing technology and create low-cost photonic devices suitable for large-scale market applications. The development of optical interconnection and silicon-based optoelectronic integrated circuit technology has a widespread application and market prospect, which will promote the development of new generation of high-performance computers, optical communication equipments and consumer electronics products.

Silicon photonics is not only useful to solve the bottleneck problem of electrical interconnection, but also can be used as an universal high-density integrated optoelectronic technology based on CMOS processing platform in microwave photonics [13, 14], quantum communication/computation [15], infrared sensing [16], graphene photonics [17], nonlinear optics [18], transformation optics [19], information photonics [20] and some novel cross-disciplines. These disciplines have also shown great vitality in recent years, and a large number of research results were published on high-impact journals. For example, in silicon-based microwave photonics, Massachusetts Institute of Technology (MIT) reported in 2013 a large-scale array antenna consisting of 4096 units, which is the silicon photonic module with the largest number of integrated components [21]. Purdue University and Chinese Academy of Sciences also realized an arbitrary waveform generator on the silicon photonic chip, which includes hundreds of various integrated units [14]. It is very challenging to achieve such a large-scale integration of so many optical units on other material platforms. In addition, the advantages of silicon-based photonics are also reflected in quantum communication. In recent two years, researchers have demonstrated two-photon interference [15] and quantum entanglement phenomenon on a silicon photonic chip. So we expect that a more complex quantum phenomenon and functional module will be realized in the near future.

1.3 Overview of Research Progress on Silicon-Based Optical Devices and Interconnections

Because of its great development and application prospects, silicon-based optical interconnection has attracted wide attention from academia and industry all over the world. The players in academia includes the University of California, Santa Barbara, Massachusetts Institute of Technology (MIT), University of Southern California, Cornell University, Purdue University, Columbia University, University of Delaware, University of Colorado, Ghent University, Universit Paris-Sud, University of Surrey, University of Southampton, Valencia Polytechnic University, Yokohama National University of Japan, University of Tokyo, Electronics and Telecommunications Eesearch Institute in South Korea, Institute of Microelectronics, Singapore, National Research Council. The players in industry not only includes Intel, IBM, Alcatel-lucent, Oracle, Cisco, Nippon Telegraph and Telephone Corp., Fujitsu Corporation, Fujikura company, Samsung and Huawei, but also some young companies like Luxtera, Kotura and Acacia.

Some countries have invested heavily to promote the rapid development of silicon based optical interconnection technology. For example, the HELIOS (pHotonics ELectronics functional Integration on CMOS) project implemented in Europe from 2008 to 2012 and including 19 European member units, aimed to integrate the optoelectronic layer and CMOS circuit by using the microelectronics process. The research contents include silicon light sources, high-speed electro-optic modulators, germanium detectors and other key optoelectronic devices, and hybrid integration technology, and optoelectronic functional devices, and CMOS monolithic integration of complex functions, so as to provide a series of solutions for the needs of industry. With a total project funding of 8 million 500 thousand euros, a total of 170 journal or conference papers were published under the support of the project. The following

PLAT4M (Photonic Libraries And Technology for Manufacturing) project has a goal of making a seamless connection between the existing silicon photonics research and industry and can adapt to different applications and production requirements. This project budget is 10 million 200 thousand euros.

From 2009 to 2013, Photonics and Electronics Technology Research Association (PETRA) in Japan implemented photonics-electronics convergence system technology (PECST) project with a goal of achieving "on-chip database", i.e., realizing some functions of databases on silicon chips.

In 2008, the Defense Advanced Research Projects Agency (DARPA) carryed out Ultra-performance Nanophotonic Intrachip Communication Program (UNIC) project with a funding of 44 million 300 thousand dollars and project members including Luxtera, Kotura, UCSD and etc. Following this project, Electronic-Photonic Heterogeneous Integration (E-PHI) was implemented with a goal of developing microelectronics technology required by micro-optoelectronic system. This compact optoelectronic system includes remote communication transceiver, coherent optical system for lidar sensor and communication equipment, optical arbitrary waveform generator and multi-wavelength imaging instrument based on comprehensive image processing and display circuit. The E-PHI project members including UCSB, Aurrion, MIT, and Delaware university. In 2011, University of Washington and Intel launched the silicon photonics development project of Optoelectronics Systems Integration in Silicon (OpSIS), and Intel participated in the form of project funding. Since 2006, IBM has launched Silicon Nanoscale-Integrated Photonic and Electronict Ransceiver (SNIPER), hoping to achieve electro-optic processor chip with a speed of 10 trillion times per second in 2018.

In recent years, China has also carried out a series of projects for silicon-based optical interconnection, which have been supported by projects such as 973 and 863. The Chinese Academy of Sciences, Peking University, Zhejiang University, Huazhong University of Science and Technology, Shanghai Jiao Tong University and other organizations have reported many innovative achievements and achieved world-class level in some research fields. But we have to know that in China, the research of silicon-based optical interconnection engineering devices is still not enough. Most of the project investment is mainly for the basic research on silicon-based photonics. Some devices are far from large-scale standardization.

Figure 1.1 shows a 4-channel wavelength division multiplexing (WDM) optical interconnection link. Starting from the laser at the transmitter, the light is modulated by an external electronic signal, and passes through a multiplexer and then enters the fiber. The receiver sends the light with different wavelengths into different photon-detectors through a demultiplexer, and then converts to a digital signal. In such an optical interconnection link, there are at least three key active devices, which are lasers, modulators and detectors. There are also a series of passive devices, such as multiplexer/demultiplexer, fiber-to-chip coupler and so on. The structure of a more complex coherent optical transceiver, is shown in Fig. 1.2. It adopts a more advanced modulation format and so there are more functional units including some polarization-controlling devices.

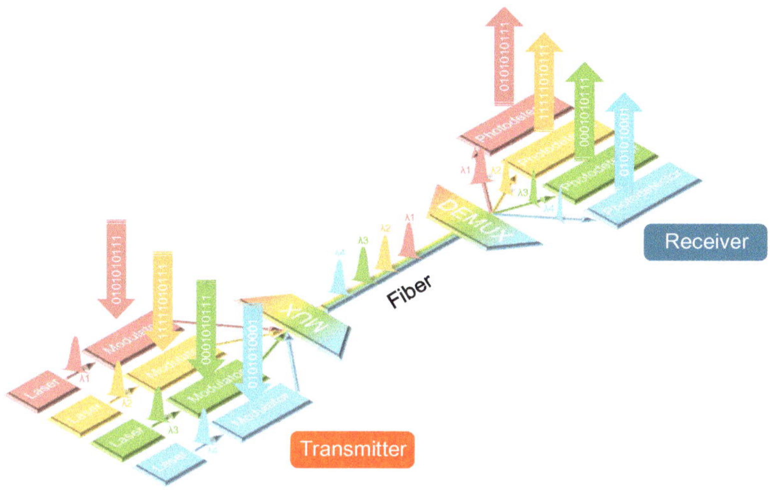

Fig. 1.1 Schematic of a 4-channel WDM silicon-based optical interconnection link

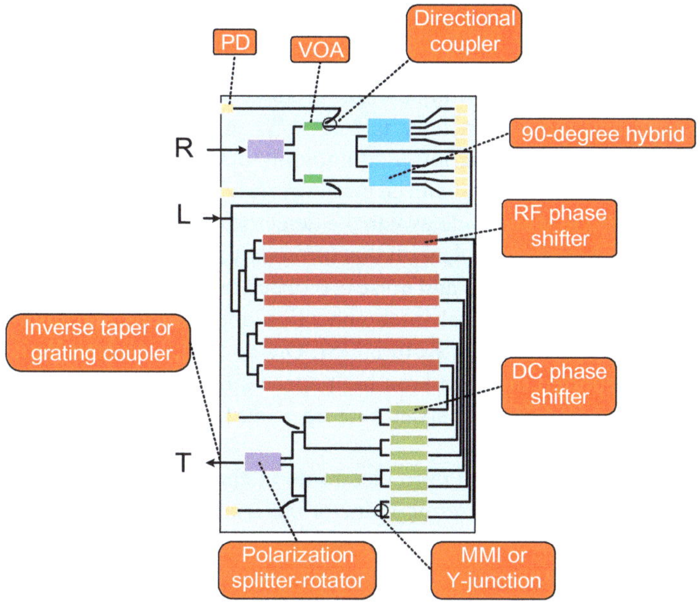

Fig. 1.2 Schematic of the structure of a silicon photonic coherent transceiver

The goal of silicon-based optical interconnection is to realize various kinds of optical functional devices on silicon. And the research of silicon based optical interconnection devices and integration technology has entered a rapid development in the past ten years.

In terms of modulator research, researchers at Intel and Cornell University reported a high speed silicon modulator based on Mach-Zehnder interferometer (MZI) [22] and micro-ring resonator (MRR) [23], respectively. Their work were published on Nature, which really opened a new era in the study of silicon photonics. After that, various silicon based modulators based on plasma dispersion effect are emerging. For a single silicon modulator, the modulation rate increases from 1 to 70 Gb/s, modulation format varies from NRZ-OOK to PDM-QPSK [24], and power consumption decreases from tens of pJ/bit to several fJ/bit [25]. These work really laid a solid foundation for high-speed, low-power-consumption and reliable electro-optic signal conversion. In terms of silicon optical modulator research, Chinese people from Chinese Academy of Sciences, Peking University and Shanghai Jiao Tong University have also reported some important work.

There are two kinds of photo-detectors, which are hybrid integrated detectors and germanium detectors. Considering germanium has a CMOS-compatible processing, it has attracted more and more attention. In 2009, Intel successfully achieved a high-speed silicon-germanium avalanche detector. This detector has achieved an unprecedented 340 GHz "gain bandwidth product". The sensitivity is −28 dB at 10 Gb/s speed and the responsivity is 5.88 A/W at 1310 nm wavelength, which makes the performance of silicon-based photonic devices is beyond the traditional optoelectronic devices in other material platforms for the first time [26]. In 2010, IBM released a 40 Gb/s Ge waveguide avalanche detector with an operation voltage of as low as 1.5 V, which is compatible with that offered by a CMOS driver. The avalanche gain can reach more than 10 dB and noise is reduced by 70%. This work is significant for various silicon-based optical interconnection applications and high-sensitivity optical receiver [27]. After that, Kotura [28], Oracle [29], Universit Paris-Sud [30], IME [31] have reported their work on high-speed silicon-germanium detector with a bandwidth up to 60 GHz and speed rate up to 40 Gb/s, which are able to satisfy various applications at present.

A difficulty of silicon-based opto-electronic integration is light source. Because silicon is an indirect-band gap semiconductor material, it is difficult to achieve light-emitting devices on silicon materials. In addition to using the coupler to couple the light of external light source into the chip, an solution is to use III-V hybrid integration. In 2006, Intel and UCSB combined the III-V materials with the SOI substrate to realize a silicon-based electrical-pumped laser for the first time [32]. In 2007, researchers at University of Ghent, Belgium, successfully developed a silicon-based hybrid micro-disk laser [33]. After that, they cascaded four lasers and coupled them into only a silicon waveguide [34]. There were a lot of reports about hybrid lasers, including distributed bragg-reflection laser [35], distributed feedback laser [36, 37] and so on, which continues to drive the hybrid integrated laser from research into practical application. Although hybrid integrated laser may be one of the most potential solutions for on-chip light sources, there have been some breakthroughs on all-silicon raman laser from Intel [38–41] and Germanium-on-silicon laser from MIT [42, 43] in the last few years. These work have provided some early technology for fully CMOS-compatible silicon based optical interconnection in the future.

In terms of passive devices, one of the typical WDM devices, silicon arrayed waveguide grating (AWG), has been demonstrated for triplexer [44], dense wavelength division multiplexing (DWDM) [45] and coarse wavelength division multiplexing (CWDM) [46]. A prototype device with 512 channels and 0.2 nm channel-spacing has been reported [47]. By optimizing the design and processing technology, University of Ghent reported an AWG device with very good performances in 2014. The device has 2 dB/1.5 dB loss and −22.5 dB/−26 dB crosstalk for 1.6 nm/3.2 nm channel spacing [48]. In addition, they used polymer-cladded narrow arrayed waveguide [49], two-dimensional grating coupler [50] and multi-mode interference coupler [51] to realize temperature- insensitive, polarization-insensitive and flat-topped optical spectrum. These work laid a solid foundation on the realization of practical on-chip multiplexing/demultiplexing. In addition, there were a lot of work about low-loss waveguide [52], etched diffraction grating (EDG) [53], mode converter [54], directional coupler (DC) [55], multi-mode interfere coupler [56], Y-junction waveguide (Y-junction) [57], optical crossing [58], 90-degree hybrid [59], grating coupler [60], inverse taper [61], polarization-controlling device [62] and mode-division multiplxer [63]. However, it is also necessary to see that there have been few papers about pure-silicon photonic devices published in high-impact journals, which indicates that the basic structure and functional categories of silicon-based optical interconnection have been stereotyped, and silicon-based optical interconnection is gradually moving from research to practicality.

The Luxtera company in the United States has led the commercialization on silicon-based optoelectronic integration. In 2006, this company launched a 4 × 10 Gb/s integrated optical transceiver module for the first time by using a standard CMOS processing. This module integrated grating coupler, MZI electro-optic modulator, germanium detector, waveguide demultiplexer, driver circuit and the receiver circuit on a single-chip [9]. In 2012, Luxtera reported its latest 4 × 28 Gb/s transciever [64]. In 2008, Intel also reported an 8 × 25 Gb/s wavelength division multiplexing module that realized 200 Gb/s data transmission [65]. In 2010, the first 4 × 12.5 Gb/s data link system integrated with a laser was launched in the world, which greatly stimulated the research of silicon photonics. In 2008, IBM proposed the concept of the silicon- based three-dimensional photonic integrated chip, and released a silicon-based optoelectronic chip in 2010. In this chip, the monolithic integration of photonic devices and electronic devices was realized [66]. After that, they implemented a 25 Gb/s monolithic integrated transceiver by using 90 nm CMOS technology [67] and a monolithic integrated 4 × 25 Gb/s transceiver chip without integrated laser was reported in May 2015. In 2012, Kotura also reported the 4 × 25 Gb/s transceiver module [68]. In 2012, several research institutions in Japan jointly reported their silicon-based optical interconnection scheme and successfully realized the transmission rate of 12.5 Gb/s per channel [69]. After that, silicon photonic transceiver module for 100 Gb/s coherent optical communication has been widely concerned in the industry. Alcatel-Lucent [70], Acacia [71], and Fujikura [72] have reported their work on it and are continuing to push the development of silicon photonic integration technology.

1.4 Overview of the Development of Silicon Photonics Industry

At present, there are not many companies that invest and develop silicon photonics in the world. There are about 30 companies in the world, which are industry leaders or start-ups. Intel is in the leading position in silicon photonics, and is currently working closely with Corning in fiber-based silicon photonics. Fujitsu, IBM, Cisco, and Huawei occupy an important position in the system field. At this stage, the industry rumors said that the market is very broad and the application is also very big from more than long-distance communication with 1000 kilometers to metro area network, from the optical access network to LAN/storage network, from device level to the back-board level, and even the inter-connections between chip and chip. However, at the present stage, silicon photonics technology is mainly used in active optical cable, 100G Ethernet and Infiniband module.

In 2008, the demand for active optical cable (AOC) in the high performance computer market began to expand due to its high durability and reliability. According to the data released by Global Information, the total sales volume of AOC was 305 thousand in 2011 with sales amount of 70 million US dollars. The company predicts that the sales will reach 786 thousand in 2016, and the sales volume will expand to 175 million US dollars. In 2008, Luxtera, which began to produce silicon photonic chip before 2008, announced that it has sold 1 million optical integrated chips with 10 Gb/s transmission capacity. Kotura, also announced in February 2013, "the sales of optical chips have doubled over the previous year, equivalent to 60 thousand channels per month." From these shipments, the two companies account for a large part of the AOC market. The reason why silicon photonics can achieve such a success in the field of AOC transceiver is that it can substantially reduce the cost through mass production, which is similar to that of semiconductor products by using CMOS technology. The previous AOC used some discrete components based on compound semiconductors, which has a higher price.

As the leader of silicon photonics technology, Intel has been acting frequently in recent years. It has maintained close cooperation with Corning, Facebook and Fujitsu in silicon photonics technology. For example, in January 2013, Intel launched AOC, which used silicon photonics technology, which supported the industry standard "Open Compute Project" led by facebook. In September 2013, Intel and Corning announced jointly a new optical fiber transmission technology which can achieve an amazing speed of 1.6 Tb/s within 820 meters. This optical fiber used ClearCurve LX multi-mode fiber technology of Corning and Intel's MXC optical interface technology. It may support Intel's silicon photonics products in the future. In November 2013, Fujitsu announced that through the great cooperation with Intel, they have successfully built and demonstrated the world's first optical PCIe server, and the key technology is Intel's painstaking research and development of silicon photonics for many years.

In addition, the market for silicon photons is also changing. The acquisition and merger emerged in an endless stream. In January 2011, Molex, an well-known interconnection company, acquired Luxtera's AOC business, including the four channel small plug and pull module (QSFP+), 40 Gb/s Ethernet and InfiniBand products. In 2013, Molex was bought by Koch for 7 billion 200 million US dollars. In February 2012, Cisco spent 271 million US dollars purchasing Lightwire and released a 100 Gb/ s optical transceiver "Cisco CPAK" based on silicon photonic technology in October. In November 2013, a 1 Tb line card using CPAK was released. In March 2013, Mellanox bought Kotura for 82 million US dollars and in the next year the founder of Kotura re-founded Rockley Photonics. In 2013, Huawei bought the Caliopa company for 7 million euros that splited from IMEC and Ghent University, and then invested its human resource and infrastructure to keep pace with its rapid growth. In December 2014, Huawei and IMEC jointly announced that they focused on optical data link technology and their strategic partnership was further enhanced.

Despite the frequent actions of silicon photonics industry in recent years, it need to be realized that the market share of silicon photons is still very small. Most of the silicon photonic products only stay in discrete devices rather than integrated chips, and the shipments are still small. On the one hand, the cost of research and development is relatively high, and the silicon photonics technology is not yet mature. On the other hand, there is still a certain gap between the silicon photonic products and the market demand. For example, the data center generally uses 850 nm and 1310 nm wavelength light source, while the silicon photonics products are working around 1550 nm. The interconnection solution by using VCSEL also poses a great challenge to silicon photonics products.

However, it need to be seen that with the increasing demand for interconnection, in addition to the advantages of silicon-based optical interconnection technology, the users and industrial chains will also promote the development of silicon photonics industry. For example, in March 2015, Microsoft and 14 data center suppliers jointly set up an organization called COBO (Consortium for On-Board Optics), aiming at establishing the standard of optical interconnection within data centers. Yole predicted that silicon photonics market will grow 38% per year, increasing from 2 million 500 thousand US dollars in 2013 to 700 million US dollars in 2024. Therefore, we have all reasons to believe that the spring of silicon photonics industry will come in the near future.

1.5 Overview of this Dissertation

In this dissertation, the author focused on engineering devices in high-speed silicon-based optical interconnections and did a lot of research on the key problems of pure silicon electro-optic modulation, advanced multiplexing technology, polarization controlling and optical waveguide coupling. I not only strived to carry out innovative research on silicon based optical interconnection, but also aimed at developing various key functional modules needed in multi-channel multiplexing transceivers and related transceivers to solve a series of problems in the early development stage of silicon photonics technology.

This dissertation was organized as followed.

1. In the second chapter, we have studied the pure silicon electro-optic modulation devices compatible with CMOS process. First, the latest development of pure light electro-optic modulator based on plasma dispersion effect is introduced. Then, the design theory of pure silicon electro-optic modulator is systematically put forward, and the method of optimizing the size of doping area in the modulator waveguide is introduced through the combination of semiconductor process simulation software and in-house programming. Then, we investigated the equivalent circuit model of carrier-depletion modulator structure with the lateral PN junction, and the extraction of various circuit elements, and optimized the traveling-wave electrode based on theory of microwave transmission line. Finally, processing and characterization of our pure silicon electro-optic modulator with two typical structures were reported.
2. In the third chapter, we study the advanced multiplexing scheme in CMOS-compatible silicon-based optical interconnection. First, we introduce the latest progress of the three most widely used multiplexing schemes, i.e., WDM, polarization multiplexing (PDM) and MDM. Then the design theory, layout drawing and experimental validation of AWG for DWDM applications were studied. Then two novel polarization beam splitter (PBS) and MDM devices for ultra-broadband applications are studied, and their performance and processing tolerance are analyzed.
3. In the fourth chapter, we have studied CMOS-compatible silicon-based polarization beam splitter (PSR). First, we introduced the research progress of silicon-based PSR, and then studied the schemes of mode conversion in the PSRs based on mode interference or mode evolution, and determines the designs of PSRs based on DC, Y-junction and MMI structures. Finally, we built an test platform for these devices and fabricated them.
4. In the fifth chapter, we study the efficient fiber-to-waveguide edge coupling mechanism. First, the coupling challenges and the latest development of the edge coupling mechanism are introduced. Then the influence of the width, length and the edge gap of the single-tip inverse taper on the coupling efficiency are studied. Then, we study the operation mechanism and performance optimization of a polarization-insensitive compact optical splitter based on the two-mode interference effect. Finally, the design theory of a novel fiber-to-waveguide edge coupler

based on double-tip inverse taper and polarization-insensitive optical splitter is studied. The relationship between various design parameters and coupling efficiency is discussed, and these couplers are processed and characterized.

References

1. Wilson L (2013) International technology roadmap for semiconductors (ITRS). Semiconductor Industry Association
2. Brian D, Arghavani R, Barlage D, Datta S, Doczy M, Kavalieros J, Murthy A, Chau R (2002) Transistor elements for 30 nm physical gate lengths and beyond. Intel Technol J 6:1–4
3. Davis JA, Venkatesan R, Kaloyeros A, Beylansky M, Souri SJ, Banerjee K, Saraswat KC, Rahman A, Reif R, Meindl JD (2001) Interconnect limits on gigascale integration (GSI) in the 21st century. Proc IEEE 89:305–324
4. Shin D, Gerstlauer A, Dömer R, Gajski DD (2005) Automatic network generation for system-on-chip communication design. In: Proceedings of 3rd IEEE/ACM/IFIP international conference on hardware/software codesign and system synthesis, pp 255–260
5. Henkel J, Wolf W, Chakradhar S (2004) On-chip networks: a scalable, communication-centric embedded system design paradigm. In: Proceedings of 17th international conference on VLSI design, pp 845–851
6. Benini L, De Micheli G (2002) Networks on chips: a new SoC paradigm. Computer 35:70–78
7. Nayfeh BA, Olukotun K (1997) A single-chip multiprocessor. Computer 30:79–85
8. Haurylau M, Chen G, Chen H, Zhang J, Nelson NA, Albonesi DH, Fauchet PM (2006) On-chip optical interconnect roadmap: challenges and critical directions. IEEE J Sel Top Quan Electron 12:1699–1705
9. Gunn C (2006) CMOS photonics for high-speed interconnects. IEEE Micro 26:58–66
10. Jalali B, Fathpour S (2006) Silicon photonics. J Lightwave Technol 24:4600–4615
11. Soref R (2006) The past, present, and future of silicon photonics. IEEE J Sel Top Quan Electron 12:1678–1687
12. Palik ED (1998) Handbook of optical constants of solids. Academic, San Diego
13. Khan MH, Shen H, Xuan Y, Zhao L, Xiao S, Leaird DE, Qi M (2010) Ultrabroad-bandwidth arbitrary radiofrequency waveform generation with a silicon photonic chip-based spectral shaper. Nat Photonics 4:117–122
14. Wang J, Shen H, Fan L, Wu R, Niu B, Varghese LT, Weiner AM (2015) Reconfigurable radio-frequency arbitrary waveforms synthesized in a silicon photonic chip. Nat Commun 6:1–8
15. Silverstone JW, Bonneau D, Ohira K, Suzuki N, Yoshida H, Iizuka N, Ezaki M, Natarajan CM, Tanner MG, Hadfield RH, Zwiller V (2014) On-chip quantum interference between silicon photon-pair sources. Nat Photonics 8:104–108
16. Soref R (2010) Mid-infrared photonics in silicon and germanium. Nat Photonics 4:495–497
17. Liu M, Yin X, Ulin-Avila E, Geng B, Zentgraf T, Ju L, Wang F, Zhang X (2011) A graphene-based broadband optical modulator. Nature 474:64–67
18. Fan L, Wang J, Varghese LT, Shen H, Niu B, Xuan Y, Weiner AM, Qi M (2012) An all-silicon passive optical diode. Science 335:447–450
19. Gabrielli LH, Liu D, Johnson SG, Lipson M (2012) On-chip transformation optics for multimode waveguide bends. Nat Commun 3
20. Vandoorne K, Mechet P, Van Vaerenbergh T, Fiers M, Morthier G, Verstraeten D, Schrauwen B, Dambre J, Bienstman P (2014) Experimental demonstration of reservoir computing on a silicon photonics chip. Nat Commun 5
21. Sun J, Timurdogan E, Yaacobi A, Hosseini ES, Watts MR (2013) Large-scale nanophotonic phased array. Nature 493:195–199

22. Liu A, Jones R, Liao L, Samara-Rubio D, Rubin D, Cohen O, Nicolaescu R, Paniccia M (2004) A high-speed silicon optical modulator based on a metal–oxide–semiconductor capacitor. Nature 427:615–618
23. Xu Q, Schmidt B, Pradhan S, Lipson M (2005) Micrometre-scale silicon electro-optic modulator. Nature 435:325–327
24. Dong P, Xie C, Chen L, Buhl LL, Chen YK (2012) 112-Gb/s monolithic PDM-QPSK modulator in silicon. Opt Express 20:B624–B629
25. Timurdogan E, Sorace-Agaskar CM, Sun J, Hosseini ES, Biberman A, Watts MR (2012) An ultralow power athermal silicon modulator. Nat Commun 5
26. Kang Y, Liu HD, Morse M, Paniccia MJ, Zadka M, Litski S, Sarid G, Pauchard A, Kuo YH, Chen HW, Zaoui WS (2009) Monolithic germanium/silicon avalanche photodiodes with 340 GHz gain–bandwidth product. Nat Photonics 3:59–63
27. Assefa S, Xia F, Vlasov YA (2010) Reinventing germanium avalanche photodetector for nanophotonic on-chip optical interconnects. Nature 464:80–84
28. Liao S, Feng NN, Feng D, Dong P, Shafiiha R, Kung CC, Liang H, Qian W, Liu Y, Fong J, Cunningham JE (2011) 36 GHz submicron silicon waveguide germanium photodetector. Opt Express 19:10967–10972
29. Li G, Luo Y, Zheng X, Masini G, Mekis A, Sahni S, Thacker H, Yao J, Shubin I, Raj K, Cunningham JE (2012) Improving CMOS-compatible Germanium photodetectors. Opt Express 20:26345–26350
30. Vivien L, Polzer A, Marris-Morini D, Osmond J, Hartmann JM, Crozat P, Cassan E, Kopp C, Zimmermann H, Fédéli JM (2012) Zero-bias 40 Gbit/s germanium waveguide photodetector on silicon. Opt Express 20:1096–1101
31. Novack A, Gould M, Yang Y, Xuan Z, Streshinsky M, Liu Y, Capellini G, Lim AEJ, Lo GQ, Baehr-Jones T, Hochberg M (2013) Germanium photodetector with 60 GHz bandwidth using inductive gain peaking. Opt Express 21:28387–28393
32. Fang AW, Park H, Cohen O, Jones R, Paniccia MJ, Bowers JE (2006) Electrically pumped hybrid AlGaInAs-silicon evanescent laser. Opt Express 14:9203–9210
33. Van Campenhout J, Rojo-Romeo P, Regreny P, Seassal C, Van Thourhout D, Verstuyft S, Di Cioccio L, Fedeli JM, Lagahe C, Baets R (2007) Electrically pumped InP-based microdisk lasers integrated with a nanophotonic silicon-on-insulator waveguide circuit. Opt Express 15:6744–6749
34. Van Campenhout J, Liu L, Romeo PR, Van Thourhout D, Seassal C, Regreny P, Di Cioccio L, Fedeli JM, Baets R (2008) A compact SOI-integrated multiwavelength laser source based on cascaded InP microdisks. Opt Express 20:1345–1347
35. Fang AW, Koch BR, Jones R, Lively E, Liang D, Kuo YH, Bowers JE (2008) A distributed Bragg reflector silicon evanescent laser. IEEE Photonics Technol Lett 20:1667–1669
36. Fang AW, Lively E, Kuo YH, Liang D, Bowers JE (2008) A distributed feedback silicon evanescent laser. Opt Express 16:4413–4419
37. Stankovic S, Jones R, Sysak MN, Heck JM, Roelkens G, Van Thourhout D (2012) Hybrid III-V/Si distributed-feedback laser based on adhesive bonding. IEEE Photonics Technol Lett 24:2155–2158
38. Rong H, Jones R, Liu A, Cohen O, Hak D, Fang A, Paniccia M (2005) A continuous-wave Raman silicon laser. Nature 433:725–728
39. Rong H, Liu A, Jones R, Cohen O, Hak D, Nicolaescu R, Fang A, Paniccia M (2005) An all-silicon Raman laser. Nature 433:292–294
40. Rong H, Xu S, Kuo YH, Sih V, Cohen O, Raday O, Paniccia M (2007) Low-threshold continuous-wave Raman silicon laser. Nat Photonics 1:232–237
41. Rong H, Xu S, Cohen O, Raday O, Lee M, Sih V, Paniccia M (2008) A cascaded silicon Raman laser. Nat Photonics 2:170–174
42. Liu J, Sun X, Camacho-Aguilera R, Kimerling LC, Michel J (2010) Ge-on-Si laser operating at room temperature. Opt Lett 35:679–681
43. Camacho-Aguilera RE, Cai Y, Patel N, Bessette JT, Romagnoli M, Kimerling LC, Michel J (2012) An electrically pumped germanium laser. Opt Express 20:11316–11320

44. Zou J, Jiang X, Xia X, Lang T, He JJ (2013) Ultra-compact birefringence-compensated arrayed waveguide grating triplexer based on silicon-on-insulator. J Lightwave Technol 31:1935–1940
45. Pathak S, Van Thourhout D, Bogaerts W (2013) Design trade-offs for silicon-on-insulator-based AWGs for (de)multiplexer applications. Opt Lett 38:2961–2964
46. Pathak S, Dumon P, Van Thourhout D, Bogaerts W (2014) Comparison of AWGs and echelle gratings for wavelength division multiplexing on silicon-on-insulator. IEEE Photonics J 6:4900109
47. Cheung S, Su T, Okamoto K, Yoo SJB (2014) Ultra-compact silicon photonic 512 × 512 25 GHz arrayed waveguide grating router. IEEE J Sel Top Quan Electron 20:310–316
48. Pathak S, Vanslembrouck M, Dumon P, Van Thourhout D, Verheyen P, Lepage G, Absil P, Bogaerts W (2014) Effect of mask discretization on performance of silicon arrayed waveguide gratings. IEEE Photonics Technol Lett 26:718–721
49. Wang L, Bogaerts W, Dumon P, Selvaraja SK, Teng J, Pathak S, Han X, Wang J, Jian X, Zhao M, Baets R (2012) Athermal arrayed waveguide gratings in silicon-on-insulator by overlaying a polymer cladding on narrowed arrayed waveguides. Appl Opt 51:1251–1256
50. Pathak S, Vanslembrouck M, Dumon P, Van Thourhout D, Bogaerts W (2012) Compact SOI-based polarization diversity wavelength de-multiplexer circuit using two symmetric AWGs. Opt Express 20:B493–B500
51. Pathak S, Vanslembrouck M, Dumon P, Van Thourhout D, Bogaerts W (2013) Optimized silicon AWG with flattened spectral response using an MMI aperture. J Lightwave Technol 31:87–93
52. Qiu C, Sheng Z, Li H, Liu W, Li L, Pang A, Wu A, Wang X, Zou S, Gan F (2014) Fabrication, characterization and loss analysis of silicon nanowaveguides. J Lightwave Technol 32:2303–2307
53. Bogaerts W, Selvaraja SK, Dumon P, Brouckaert J, De Vos K, Van Thourhout D, Baets R (2010) Silicon-on-insulator spectral filters fabricated with CMOS technology. IEEE J Sel Top Quan Electron 16:33–44
54. Zou J, Yu Y, Ye M, Liu L, Deng S, Xu X, Zhang X (2014) Short and efficient mode-size converter designed by segmented-stepwise method. Opt Lett 39:6273–6276
55. Mikkelsen JC, Sacher WD, Poon JK (2014) Dimensional variation tolerant silicon-on-insulator directional couplers. Opt Express 22:3145–3150
56. Sheng Z, Wang Z, Qiu C, Li L, Pang A, Wu A, Wang X, Zou S, Gan F (2012) A compact and low-loss MMI coupler fabricated with CMOS technology. IEEE Photonics J 4:2272–2277
57. Zhang Y, Yang S, Lim AEJ, Lo GQ, Galland C, Baehr-Jones T, Hochberg M (2013) A compact and low loss Y-junction for submicron silicon waveguide. Opt Express 21:1310–1316
58. Liu Y, Shainline JM, Zeng X, Popovi MA (2014) Ultra-low-loss CMOS-compatible waveguide crossing arrays based on multimode Bloch waves and imaginary coupling. Opt Lett 39:335–338
59. Yang W, Yin M, Li Y, Wang X, Wang Z (2013) Ultra-compact optical 90-degree hybrid based on a wedge-shaped 2 × 4 MMI coupler and a 2 × 2 MMI coupler in silicon-on-insulator. Opt Express 21:28423–28431
60. Zaoui WS, Kunze A, Vogel W, Berroth M, Butschke J, Letzkus F, Burghartz J (2014) Bridging the gap between optical fibers and silicon photonic integrated circuits. Opt Express 22:1277–1286
61. Cardenas J, Poitras CB, Luke K, Luo LW, Morton PA, Lipson M (2014) High coupling efficiency etched facet tapers in silicon waveguides. IEEE Photonics Technol Lett 26:2380–2382
62. Dai D, Bauters J, Bowers JE (2012) Passive technologies for future large-scale photonic integrated circuits on silicon: polarization handling, light non-reciprocity and loss reduction. Light Sci Appl 1:1–12
63. Luo LW, Ophir N, Chen CP, Gabrielli LH, Poitras CB, Bergmen K, Lipson M (2014) WDM-compatible mode-division multiplexing on a silicon chip. Nat Commun 5:1–7
64. Mekis A, Abdalla S, De Dobbelaere PM, Foltz D, Gloeckner S, Hovey S, Jackson S, Liang Y, Mack M, Masini G, Novais R (2012) Scaling CMOS photonics transceivers beyond 100 Gb/s. In: SPIE OPTO, p 82650A

65. Liu A, Liao L, Chetrit Y, Basak J, Nguyen H, Rubin D, Paniccia M (2008) 200 Gbps photonic integrated chip on silicon platform. In: IEEE 5th international conference on group IV photon, pp 368–370
66. Green W, Assefa S, Rylyakov A, Schow C, Horst F, Vlasov Y (2010) CMOS integrated silicon nanophotonics: enabling technology for exascale computational systems. In: SEMICON, Chiba, Japan, pp 1–3
67. Assefa S, Shank S, Green W, Khater M, Kiewra E, Reinholm C, Kamlapurkar S, Rylyakov A, Schow C, Horst F, Pan H (2012) A 90 nm CMOS integrated nano-photonics technology for 25 Gbps WDM optical communications applications. In: IEEE international electron devices meeting (IEDM), vol 33, pp 31–33
68. Jatar S, Malinge Y, Zhou Z, Liang H, Liao S, Li Z, Bushyakanist C, Lee DC, Shafiiha R, Luff J, Feng D (2012) Performance of parallel 4 $\times$ 25 Gbs transmitter and receiver fabricated on SOI platform. In: IEEE 9th international conference on group IV photon (GFP), pp 159–161
69. Urino Y, Noguchi Y, Noguchi M, Imai M, Yamagishi M, Saitou S, Hirayama N, Takahashi M, Takahashi H, Saito E, Okano M (2012) Demonstration of 12.5-Gbps optical interconnects integrated with lasers, optical splitters, optical modulators and photodetectors on a single silicon substrate. Opt Express 20:B256–B263
70. Dong P, Liu X, Chandrasekhar S, Buhl LL, Aroca R, Chen YK (2014) Monolithic silicon photonic integrated circuits for compact 100+Gb/s coherent optical receivers and transmitters. IEEE J Sel Top Quan Electron 20:150–157
71. Doerr CR, Chen L, Vermeulen D, Nielsen T, Azemati S, Stulz S, McBrien G, Xu XM, Mikkelsen B, Givehchi M, Rasmussen C (2014) Single-chip silicon photonics 100-Gb/s coherent transceiver. In: Optical fiber communication conference, pp Th5C–1
72. Milivojevic B, Raabe C, Shastri A, Webster M, Metz P, Sunder S, Chattin B, Wiese S, Dama B,Shastri K (2013) 112 Gb/s DP-QPSK transmission over 2427 km SSMF using small-size silicon photonic IQ modulator and low-power CMOS driver. In: Optical fiber communication conference, p OTh1D–1

Chapter 2
CMOS-Compatible Silicon Electro-Optic Modulator

Abstract In this chapter, we have studied the pure silicon electro-optic modulation devices compatible with CMOS process. First, the latest development of pure electro-optic modulator based on plasma dispersion effect is introduced. Then, the design theory of pure silicon electro-optic modulator is systematically put forward, and the method of optimizing the size of doping area in the modulator waveguide is introduced through the combination of semiconductor process simulation software and in-house programming. Then, we investigated the equivalent circuit model of carrier-depletion modulator structure with the lateral PN junction, and the extraction of various circuit elements, and optimized the traveling-wave electrode based on theory of microwave transmission line. Finally, processing and characterization of our pure silicon electro-optic modulator with two typical structures were reported.

2.1 Principle of Silicon Electro-Optic Modulator: Plasma Dispersion Effect

The plasma dispersion effect describes the change of the refractive index and optical absorption coefficient caused by the free carrier in the semiconductor. In this section we first derive the plasma dispersion effect, which is referred from Dr. Fuwan Gan's dissertation [1].

The motion of an electron in silicon can be described by the Newton's laws.

$$\mathbf{F}_{tot} = m\frac{d\mathbf{r}}{dt} = \mathbf{F}_{ext} - \mathbf{F}_{damp} \tag{2.1}$$

Here $\mathbf{r}$ and $\mathbf{F}_{tot}$ represent the position and force of the electron, respectively. $\mathbf{F}_{ext} = q(\mathbf{E} + \mathbf{v} \times \mathbf{B}) \simeq q\mathbf{E}$ represents the external force of the electron. When $\mathrm{v} \ll \mathrm{c}_l$, the second term can be ignored. Here c_l is the velocity of the plane wave and the resistance $\mathbf{F}_{damp}$ can be extracted from the change of the momentum in the band.

J. Wang, *CMOS-Compatible Key Engineering Devices for High-Speed Silicon-Based Optical Interconnections*, Springer Theses,
https://doi.org/10.1007/978-981-13-3378-1_2

$$\mathbf{F}_{damp} = \frac{m\mathbf{v}}{\tau} \tag{2.2}$$

Here τ represents the relaxation time constant. And the motion equation of the electron can be rewritten as

$$q\mathbf{E} = m\frac{d\mathbf{v}}{dt} + \frac{m\mathbf{v}}{\tau} \tag{2.3}$$

If we suppose that the electrical field $\mathbf{E} = \mathbf{E}_0 e^{-\omega t}$, then the electron velocity can be expressed as

$$\mathbf{v} = \frac{q\mathbf{E}}{-i\omega m + m/\tau} \tag{2.4}$$

We put the electron velocity into Maxwell's equations and then

$$\begin{aligned} \nabla \times \mathbf{H} &= \frac{\partial \mathbf{D}}{\partial t} + \mathbf{J}_p \\ &= -i\omega\epsilon\mathbf{E} + Nq\mathbf{v} \\ &= -i\omega\epsilon_0[\epsilon_r + i\epsilon_i - \frac{(\omega_p\tau)^2(\omega\tau)^2 - i\omega\tau}{(\omega\tau)^2(\omega\tau)^2 + 1}]\mathbf{E} \\ &= -i\omega\epsilon_0\epsilon\mathbf{E} \end{aligned} \tag{2.5}$$

and

$$\epsilon \equiv (n + ik)^2 = \epsilon_r + i\epsilon_i - \frac{(\omega_p\tau)^2(\omega\tau)^2 - i\omega\tau}{(\omega\tau)^2(\omega\tau)^2 + 1} \tag{2.6}$$

Here ω_p is the plasmon frequency of the electron, $\mathbf{J}_p$ is the polarization current density, N is the free-carrier density and m is the effective mass of the carrier. ϵ_r and ϵ_i are the real and imaginary part of the relative permittivity of a dielectric without polarization current, respectively. ϵ_0 and $\epsilon = (n + ik)^2$ are the vacuum dielectric constant and the effective dielectric constant, respectively. So we can obtain the relationship between the real part of the refractive index and the imaginary part k.

$$n^2 - k^2 = \epsilon_r - \frac{(\omega_p\tau)^2}{(\omega\tau)^2 + 1} \tag{2.7}$$

$$2nk = \epsilon_i + \frac{(\omega\tau)^2}{\omega\tau}\frac{1}{(\omega\tau)^2 + 1} \tag{2.8}$$

If we let the total force $\mathbf{F}_{tot}$ zero, then we can get the electron mobility $\mu = q\tau/m$ according to Eqs. 2.1 and 2.4. We put $\tau = m\mu/q$, n = $n_0+\Delta n$ and k = $k_0+\Delta k$ into Eqs. 2.7 and 2.8, then we can get the change of the real part of the refractive index and the change of the absorption coefficient as

$$\Delta n = \frac{e^2\lambda^2}{8\pi^2 c^2 \epsilon_0 n_0}\frac{N}{m} \tag{2.9}$$

$$\Delta\alpha = \frac{4\pi\,\Delta k}{\lambda} = \frac{e^3\lambda^2}{4\pi^2 c^3 \epsilon_0 n_0}\frac{N}{m^2\mu} \tag{2.10}$$

Here $\omega\tau \gg 1$, $n \gg k$ and $n_0 \gg \Delta n$ can be satisfied in the infrared frequency. n_0 and k_0 are the real and imaginary part of the refractive index without a free carrier, respectively. Considering the effect of electrons and holes, the total change of the refractive index is

$$\Delta n = -\frac{e^2\lambda^2}{8\pi^2 c^2 \epsilon_0 n}\left(\frac{\Delta n_e}{m_e} + \frac{\Delta n_h}{m_h}\right) \tag{2.11}$$

The change of the absorption coefficient is

$$\Delta\alpha = \frac{4\pi\,\Delta k}{\lambda} = \frac{e^3\lambda^2}{4\pi^2 c^3 \epsilon_0 n}\left(\frac{\Delta n_e}{m_e^2\mu_e} + \frac{\Delta n_h}{m_h^2\mu_h}\right) \tag{2.12}$$

Here Δn_e and Δn_h are the electron density and hole density, respectively. Equations 2.11 and 2.12 are the *relationship between the refractive index and absorption coefficient and the carrier concentration in semiconductor materials, respectively.* With the increase of the carrier concentration in the semiconductor material, the refractive index will gradually decrease and the absorption of the light field will be gradually enhanced. Therefore, by changing the concentration of the carrier, the absorption coefficient of the material will be changed while the refractive index of the material is changed. Finally, the phase and intensity of light can be changed by the change of the refractive index and absorption and thus the modulation of the light wave can be realized.

Another method of predicting the change of refractive index is based on the experimental data of electrical absorption by using the Kramers-Kronig relationship.

$$\Delta n(\omega) = \frac{c}{\pi} P \int_0^\infty \frac{\Delta\alpha(\omega')d\omega'}{\omega'^2 - \omega^2} = 6.3 \times 10^{-6} P \int_0^\infty \frac{\Delta\alpha(V')dV'}{V'^2 - V^2} \tag{2.13}$$

Here ω is the angular frequency, $V = \hbar\omega/q$ is the normalized photon energy and c is the light velocity. By using the K-K relationship, Soref et. al., collected the absorption spectrum of some highly-doped silicon samples and deduced a more direct relationship [2]. When the wavelength is 1.3 μm,

$$\Delta n = -6.2 \times 10^{-22}\Delta n_e - 6.0 \times 10^{-18}(\Delta n_h)^{0.8} \tag{2.14}$$

$$\Delta\alpha = 6.0 \times 10^{-18}\Delta n_e + 4.0 \times 10^{-18}\Delta n_h \tag{2.15}$$

When the wavelength is 1.55 μm,

$$\Delta n = -8.8 \times 10^{-22} \Delta n_e - 8.5 \times 10^{-18} (\Delta n_h)^{0.8} \tag{2.16}$$

$$\Delta \alpha = 8.5 \times 10^{-18} \Delta n_e + 6.0 \times 10^{-18} \Delta n_h \tag{2.17}$$

Here Eqs. 2.16 and 2.17 are the principles of the silicon electro-optical modulator based on plasma dispersion effect.

2.2 Research Progress of Silicon Electro-Optic Modulator

2.2.1 Summary

As a result of the joint efforts of the academic and industrial community [3, 4], the silicon-based electro-optic modulator has made great progress in the last ten years [5, 6]. For most data transmission applications, the optical modulator is at the heart of the opto-electronic system. The silicon-based electro-optic modulator based on silicon photonics not only provides a low-cost solution for optical communication, but also makes it easier to achieve short-distance optical interconnection.

Although there is no linear electro-optic effect (Pockel effect) in silicon material, the use of other modulation mechanism can also achieve modulation devices, such as the Pockel effect caused by strain [7], Kerr effect [8], Franz-Keldysh effect [9], thermo-optic effect, QCSE (Quantum-confined Stark effect) [10] and plasma dispersion effect [2].

Moreover, silicon can also be combined with other materials. A series of hybrid silicon-based or silicon-compatible modulation devices, such as III-V [11–16], germanium [17–21], polymer [8, 22–25] and graphene [26], have been realized. Especially in the near future, plasmonics has been introduced into the silicon-based electro-optic modulator and a series of novel devices have been obtained [27].

Figure 2.1 shows the operation principle of a optical modulator. When the input light is continuous, optical modulator can transform information carried by electric signal to optical signal. The change of current or voltage corresponds to the intensity change of the optical signal.

In the last section, we introduced the operation basis of silicon optical modulator: plasma dispersion effect. In order to change the carrier concentration in the waveguide and to realize the interaction between carrier and light controlled by the voltage or current, there are three mechanisms. They are carrier accumulation effect, injection effect and depletion effect respectively, as shown in Fig. 2.2a–c.

For the injection type modulator, the carrier-injection structure is usually realized by the PIN junction when the waveguide is mainly intrinsic silicon. When positive bias is applied to the device, the electrons and holes will be injected into the waveguide region of the device, resulting in the decrease of the refractive index and changing the

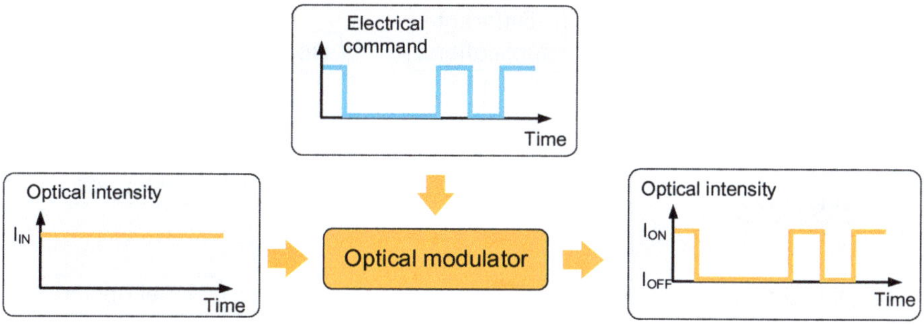

Fig. 2.1 Principle of optical modulator

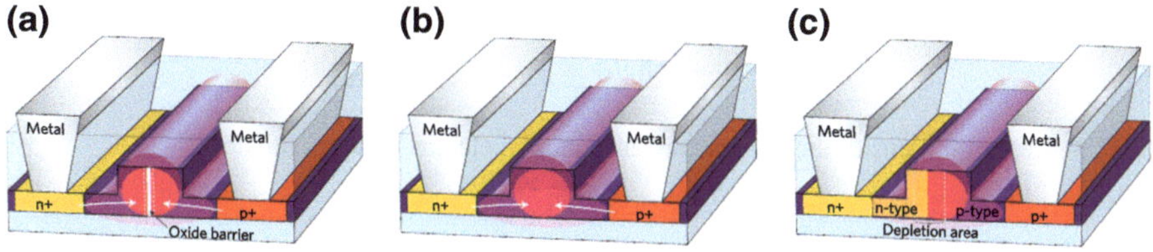

Fig. 2.2 Three kinds of silicon electro-optic modulators based on plasma dispersion effect. **a** Accumulation type. **b** Injection type. **c** Depletion type. Reprinted by permission from Macmillan Publishers Ltd: Ref. [6], copyright 2010

phase of light passing through the device. In the early stage of the silicon photonics, the modulator is mainly based on the carrier injection effect. Such a modulator was suitable for millimeter-size waveguide in that era, which reduced the difficulty of the process, and ensured the efficient coupling between waveguide and optical fiber. In the middle of the 1980s, Soref and Bennett reported *the first modulator based on the plasma dispersion effect* [28]. Although the modulator has made great progress in the past decades, its modulation bandwidth is limited by the relatively long lifetime of the minority carrier in silicon material, so it usually only has the bandwidth of MHz. It was once proposed to overcome the shortcoming by using the method of signal pre-emphasis. And a series of high-speed modulation devices were reported. For example, in 2007, Manipatruni et. al., at Cornell University proved 18 Gb/s high speed transmission [29]. However, we must also see that the complexity of the driving circuit is also increasing. And with the further improvement of the modulation speed, the driving circuit is likely to become a new problem.

By using the accumulating effect of the carrier, limitation on the modulator bandwidth can be avoided by the carrier lifetime. A carrier accumulating modulator uses an insulating layer in the middle of the waveguide, so when the bias voltage is applied to the device, the free carrier will accumulate on both sides of the insulation layer, just like a dielectric layer at both ends of the capacitor. Because the change of the carrier concentration at this time is mainly in the insulating layer, it is more efficient than the

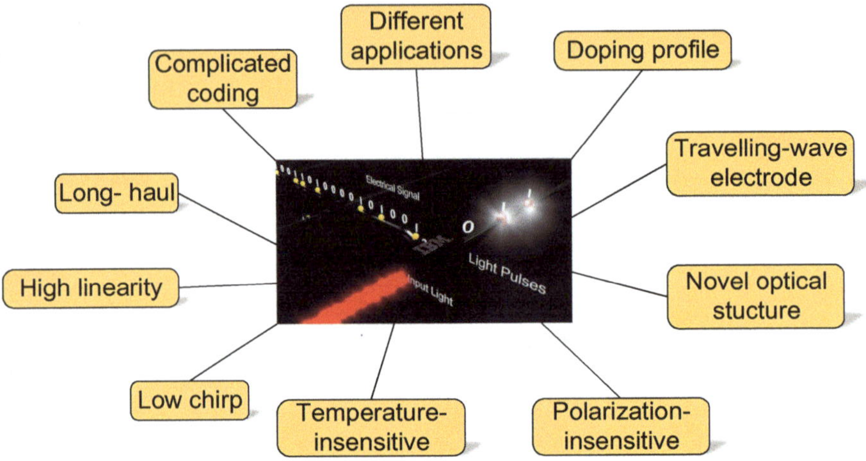

Fig. 2.3 Research focus of the silicon carrier-depletion electro-optic modulator

injection type. And as the size of the device goes down further, it will become more advantageous. In 2004, Intel's first silicon optical modulator, which has more than GHz frequency, was based on the accumulation effect [30]. However, the structure of this kind of modulator is a little complex and requires a higher level of process, such as the need for oxidation of a insulator layer in the silicon waveguide.

The bandwidth of a carrier depletion modulator is also not limited by the lifetime of the minority carrier. This modulator placed a PN junction in the waveguide to make the carrier interact with the light and reduce the carrier concentration by applying the reverse bias voltage and increasing the width of the depletion region. Similar to the accumulation modulator, the change of carrier concentration is also limited in the stronger mode field, so it is also suitable for small waveguides. In this chapter, we will discuss the silicon electro-optic modulator more specifically.

In 2005, Gardes proposed the structure of carrier-depletion silicon optical modulator for the first time and predicted its intrinsic bandwidth up to 50 GHz [31]. In 2007, a modulator with similar structure was proved by Liu et. al., from Intel. It can provide up to 30–40 Gb/s modulation rate [32, 33]. After that, the carrier-depletion silicon optical modulator has been widely concerned. We have briefly summarized the research focus of the carrier-depletion silicon electro-optic modulator, as shown in Fig. 2.3. The hot topics of research include not only the optimization and improvement of modulator, but also its applications in long-distance optical communication, analog communication, microwave photonics and mid-infrared sensing. Therefore, it is necessary for us to study and summarize the hot spots.

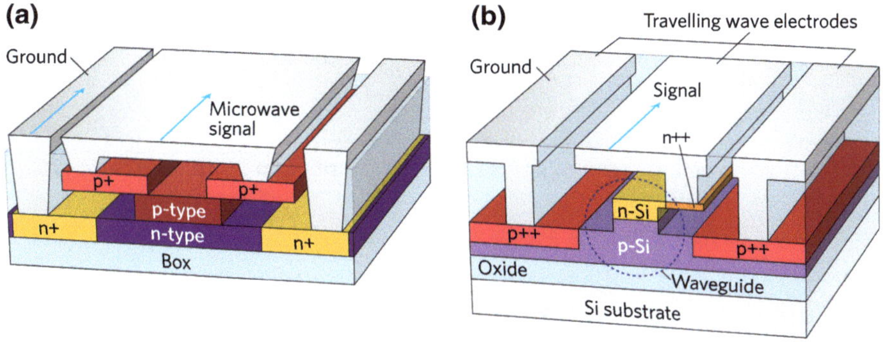

Fig. 2.4 Schematic of the cross-section of the silicon carrier-depletion electro-optic modulator based on the vertical PN junction. **a** First theoretical model [31]. **b** schematic of the device cross-section realized by Intel [32]. Reprinted by permission from Macmillan Publishers Ltd: Ref. [6], copyright 2010

2.2.2 Optimization of Doping Region

The optimization of the waveguide doping region is mainly to improve the modulation efficiency and reduce the absorption loss. Although there have been many work on doping region design in the recent ten years, the structure of doping regions can be classified into three categories according to the direction of PN junction. Figure 2.4 shows the design of PN junction with vertically-doped region. This kind of PN junction structure has attracted much attention in the early stage. For example, the first carrier-depletion silicon optical modulator was based on this structure [31–33].

A more common structure is based on a horizontal PN junction, as shown in Fig. 2.5a. The modulator with this doped structure was first realized in 2008. Most of the researchers chose the structure with the same width of P and N areas, namely offset = 0, and the optimization of performance is mainly achieved by optimizing waveguide sizes, doping concentrations and other parameters. Although the modulation rate of 40–50 Gb/s has been reported [34, 35], it is obvious that the offset is also an important design parameter because the effect of electron and hole changes on the refractive index is different according to the plasma dispersion effect [36–39]. Because the P-type doping can change more refractive index, the region of the waveguide is larger in the P-type region. The design optimization on this type of doping areas can be seen in the subsequent chapter and we will discuss it in detail.

We can notice that the change of refractive index is mainly achieved through the change of the width of the depletion region, while the doped area outside the depletion region has little effect on the refractive index change. However, this part of the doped area increases the absorption loss. So some researchers put forward some new structures to ensure good modulation efficiency while reducing light absorption loss. A typical structure, as shown in Fig. 2.5b, was proposed by Marris-Morini. In the middle of the waveguide, there is only a limited width of P doped region, while

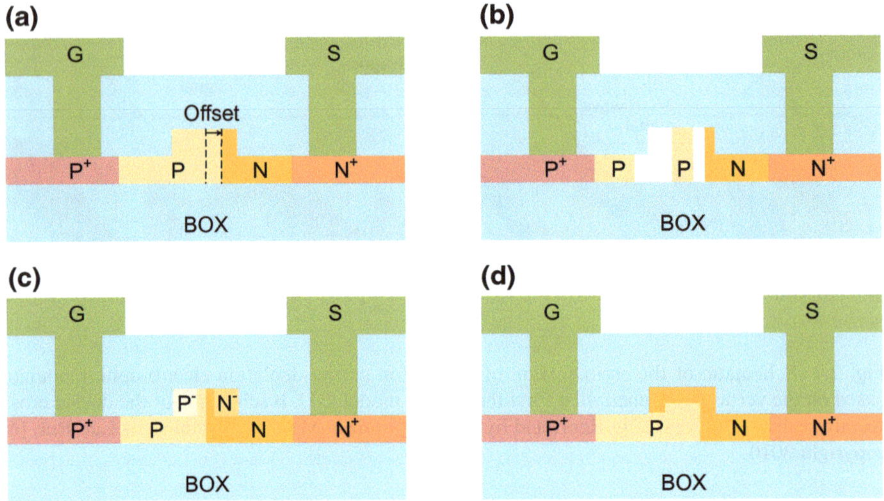

Fig. 2.5 Schematic of the cross-section of the silicon carrier-depletion electro-optic modulator based on the horizontal PN junction

most of the waveguide region is undoped. When the bias is applied, no depletion region is formed at the edge of the waveguide, so no additional absorption loss is caused. Marris-Morini et. al., have done a series of systematic work based on such PIPIN structures, including device modeling [40, 41], MZI modulator [42, 43], MRR-assisted MZI modulator [44] and polarization-insensitive modulator [45]. In 2013, 40 Gb/s modulation speed was proved [46]. Tu et. al., from IME also proposed a way of doping compensation to reduce the doping concentration of the two corners in the ridge waveguide, which thereby reduced the absorption loss [47–49]. High-speed transmission of 50 Gb/s was proved [48] and the structure is shown in Fig. 2.5c.

Another problem that needs to be taken into consideration in the design of horizontal PN junction is fabrication tolerance. Because the device performance is quite sensitive to the offset in Fig. 2.5a, we must achieve a high-tolerant PN junction and its corresponding processing technology, so as to increase the yield. Gardes and Thomson et. al., from Surrey University proposed a self-alignment technology to realize the structure as shown in Fig. 2.5d [35, 50–53]. And in a 3.5 mm-long modulator, 40 Gb/s transmission speed with 10 dB extinction ratio was proved [53] while 50 Gb/s transmission with 3 dB extinction ratio was demonstrated in a 1-mm long modulator [35].

Another kind of PN junction structure is formed in the direction of light propagation, as shown in Fig. 2.6a. Because when the length of the cycle is small the depletion region of this interdigitated structure is larger than that of the other two kinds of PN junction structures, it is possible to get a particularly good modulation efficiency. Another advantage is that its process tolerance is better. However, the increase in the depletion region will result in increasing on the capacitance in the

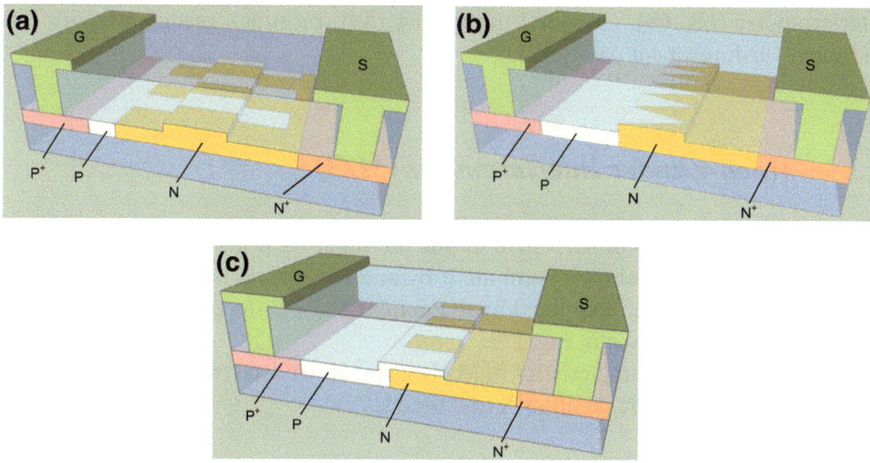

Fig. 2.6 Schematic of the cross-section of the silicon carrier-depletion electro-optic modulator based on the interdigital PN junction

unit length, which has a negative impact on the modulation speed and bandwidth performance. Such a structure was first put forward by Li et. al., in 2009 [54], and a lot of high-speed modulators with diferent optical structures were reported in the following years with highest speed of 44 Gb/s [55–59]. The researchers at the University of Ghent also compared the performance of the MZI/MRR modulators with lateral/interdigitated PN junctions [60, 61]. Ziebell et. al., have also implemented a 10 Gb/s MRR modulator by using similar structures [62].

By combining PN junctions in all directions, the PN junction structure combined in two or three directions can be realized. Figure 2.6b shows a zigzag PN junction consisting of a horizontal PN junction and a interdigitated PN junction [63]. The test results show that the modulation efficiency of the modulator is 1.7 V·cm, and the high-speed transmission of 44 Gb/s can be obtained. Such a structure was also proposed by Spector from MIT [64] and proved by the Liow from IME in Singapore [65]. The structure in Fig. 2.6d can also be seen as a mixture of the horizontal PN junction and the vertical PN junction [51]. Furthermore, it is a combination of PN junction in three directions, as shown in Fig. 2.6c. Resenberg et. al., in IBM have implemented such a structure and proved the high-speed transmission of 40 Gb/s [66].

2.3 Optical Design of Silicon Electro-Optic Modulator

The optical structure of silicon electro-optic modulator mainly includes Mach-Zhnder interferometer (MZI) and micro-ring resonator (MRR), which have been widely studied. In addition, people proposed MRR-assisted MZI [67], Michelson interferometer [68, 69], cascaded micro-ring [55], racetrack resonant cavity [70] and micro-disk resonant cavity [71].

2.3.1 Mach-Zhnder Interferometer (MZI)

The operation principle of MZI is to divide the input light into two beams through a light beam splitter, and then combine it by another optical beam combiner. Then the output signal can be modulated by the optical phase difference of two beams. A typical MZI can be shown in Fig. 2.7.

The transmission characteristics of MZI can be derived as follows. First, the input field can be expressed by position X and time t.

$$E_i(x,t) = E_0 e^{-j(kx-\omega t)} \tag{2.18}$$

Here E_0, k, and ω are electrical intensity, wave number, and angular frequency, respectively. If light is divided into two beams with the same intensity in the beam splitter, the amplitude of the electric field of each beam is

$$E_1 = E_2 = \frac{1}{\sqrt{2}} E_0 \tag{2.19}$$

At the end of each arm of the MZI, the amplitude of the electric field can be written as

$$E_1' = E_1 e^{-j\phi_1 - \alpha L_1} = \frac{1}{\sqrt{2}} E_0 e^{-j\phi_1 - \alpha L_1} \tag{2.20}$$

$$E_2' = E_2 e^{-j\phi_2 - \alpha L_2 - j\phi_B} = \frac{1}{\sqrt{2}} E_0 e^{-j\phi_2 - \alpha L_2 - j\phi_B} \tag{2.21}$$

Here ϕ_1 and ϕ_2 are the phase change of the electric field in two arms, respectively. αL_1 and αL_2 are the attenuation of electric field in two arms, respectively. And ϕ_B is the additional phase difference between the two arms at the offset point. So the total output electric field of the MZI can be written as

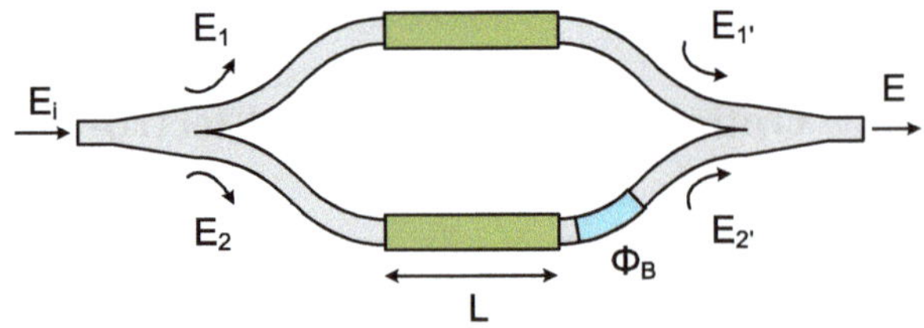

Fig. 2.7 Schematic of Mach-Zehner interferometer for optical modulator

$$\begin{aligned} E &= \frac{1}{\sqrt{2}}(E_1' + E_2') \\ &= \frac{1}{\sqrt{2}} E_0 e^{-j\phi_{DC} - \alpha L_{DC}} (e^{-j\Delta\phi - \Delta\alpha L} + e^{j\Delta\phi + \Delta\alpha L - j\phi_B}) \\ &= \frac{1}{\sqrt{2}} E_0 e^{-j\phi_{DC} - \alpha L_{DC} - j\frac{\phi_B}{2}} (e^{-j\Delta\phi - \Delta\alpha L + j\frac{\phi_B}{2}} + e^{j\Delta\phi + \Delta\alpha L - j\frac{\phi_B}{2}}) \end{aligned} \tag{2.22}$$

Here $\phi_{DC} \equiv (\phi_1 + \phi_2)/2$, $\alpha L_{DC} \equiv (\alpha L_1 + \alpha L_2)/2$, $\Delta\phi \equiv (\phi_1 - \phi_2)/2$ and $\Delta\alpha L \equiv (\alpha L_1 - \alpha L_2)/2$. If the modulator works in a push-pull mode, ϕ_{DC} and αL_{DC} are constants. The normalized power transmission characteristics can be written as

$$\begin{aligned} T &= \frac{E \cdot E^*}{E_i \cdot E_i^*} = \frac{1}{2E_0^2} E \cdot E^* \\ &= \frac{1}{4} e^{-2\alpha L_{DC}} [e^{-2\Delta\alpha L} + e^{2\Delta\alpha L} + 2\cos(2\Delta\phi - \phi_B)] \\ &= \frac{1}{2} e^{-2\alpha L_{DC}} [\cosh(2\Delta\alpha L) + \cos(2\Delta\phi - \phi_B)] \\ &\simeq \frac{1}{2} e^{-2\alpha L_{DC}} [\cosh(\frac{\Delta\phi}{4}) + \cos(2\Delta\phi - \phi_B)] \end{aligned} \tag{2.23}$$

Here $2\Delta\phi = \frac{2\pi}{\lambda}\Delta n L$ and $2\Delta\alpha L = \frac{\Delta\phi\lambda}{2\pi\cdot\mu m} \simeq \Delta\phi/4$ are used to simply the expression. The approximate relation at wavelength $\lambda = 1.55\mu m$ is established if only the electron in the plasma dispersion effect is considered.

We extract the first-order derivative and second-order derivative of the output light intensity on $\Delta\phi$ as follows

$$\begin{aligned} \frac{\partial T}{\partial \Delta\phi}|_{\Delta\phi=0} &= \frac{1}{2} e^{-2\alpha L_{DC}} \left[\frac{1}{4}\sinh(\frac{\Delta\phi}{4}) - 2\sin(2\Delta\phi - \phi_B)\right] \\ &= e^{-2\alpha L_{DC}} \sin(\phi_B) \end{aligned} \tag{2.24}$$

$$\begin{aligned} \frac{\partial^2 T}{\partial \Delta\phi^2}|_{\Delta\phi=0} &= \frac{1}{2} e^{-2\alpha L_{DC}} [\frac{1}{16}\cosh(\frac{\Delta\phi}{4}) - 4\cos(2\Delta\phi - \phi_B)] \\ &= \frac{1}{2} e^{-2\alpha L_{DC}} [\frac{1}{16} - 4\cos(\phi_B)] = 0 \end{aligned} \tag{2.25}$$

In order to remove the second-order parameters, the second-order derivative is 0 as Eq. 2.25. So we can get

$$\phi_B = \cos^{-1}(\frac{1}{4 \times 16}) \simeq 89.1^\circ \tag{2.26}$$

In addition, in the MZI modulator, if the length of the two arms in the modulation region are same, i. e., $L_1 = L_2 = L$ and $\alpha L_{DC} = 0$, Eq. 2.23 can be simplified as

$$\begin{aligned} T &= \frac{1}{2}e^{-2\alpha L_{DC}}[\cosh(2\Delta\alpha L) + \cos(2\Delta\phi - \phi_B)] \\ &= \frac{1}{2}e^{-2\alpha L_{DC}}[1 + \cos(2\Delta\phi - \phi_B)] \\ &= e^{-2\alpha L}\cos^2(\Delta\phi - \frac{\phi_B}{2}) \end{aligned} \tag{2.27}$$

Equation 2.27 presents the transmission characteristics of an asymmetric MZI designed with the same electro-optic modulation region. We can see constructive interference or destructive interference occur at the output by tuning $\Delta\phi$ and ϕ_B.

Here, we only consider a general MZI. In fact, the phase performance of MZI is influenced by different optical splitter and combiner.

2.3.2 *Micro-ring Resonator (MRR)*

The MRR structure most commonly used in modulator design is composed of a micro-ring resonator with a radius of R and a waveguide, as shown in Fig. 2.8. Here we derive the transmission relationship in this structure, and the main reference of this section is Ref. [72].

Suppose that the mode excited in this MRR is unidirectional and the coupling is nondestructive. So all kinds of loss in the propagation can be uniformly represented by the attenuation coefficient α. We only consider one polarization state and the MRR does not produce or couple other polarization states. Then the relationship between the input and output electric fields of the MRR can be represented by the transfer matrix.

$$\begin{pmatrix} E_{t1} \\ E_{t2} \end{pmatrix} = \begin{pmatrix} t & \kappa \\ -\kappa^* & t^* \end{pmatrix} \begin{pmatrix} E_{i1} \\ E_{i2} \end{pmatrix} \tag{2.28}$$

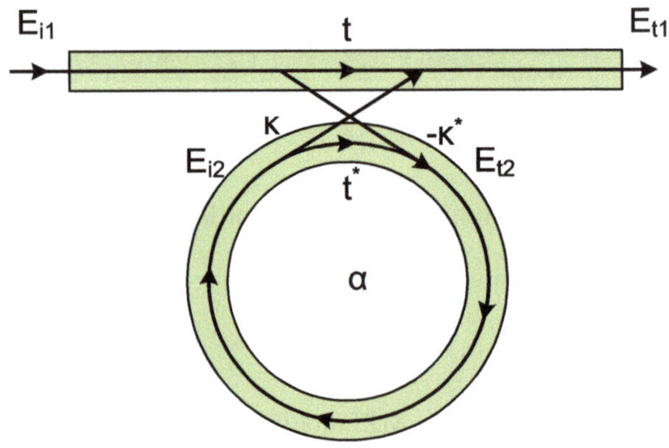

Fig. 2.8 Schematic of micro-ring resonator for modulator

The amplitude of the E's complex mode field has been normalized, so the square of it corresponds to the mode power. t and κ are the amplitude transmission coefficient and the amplitude coupling coefficient, respectively. Because of the reciprocity of light, the matrix is symmetric, corresponding to

$$|\kappa|^2 + |t|^2 = 1 \tag{2.29}$$

In order to simplify the model, E_{i1} can be assumed to be 1, then E_{i2} can be represented as

$$E_{i2} = \alpha \cdot e^{j\theta} E_{t2} \tag{2.30}$$

Here α is the attenuation coefficient of micro-ring ($\alpha = 1$ represents lossless), $\theta = \omega L/c$, $L = 2\pi r$ is the circumference of micro-ring, r is the distance between center of the circle and the waveguide, $c = c_o/n_{eff}$ is the phase velocity of the mode in micro-ring, the angular velocity $\omega = kc_o$, c_o is the light speed in vacuum. The wave-number k in vacuum satisfies the relationship with the wavelength λ: $k = 2\pi/\lambda$. By using the vacuum wave number k, the effective refractive index n_{eff} of the model can be introduced into the coupling relationship of the micro-ring.

$$\beta = k \cdot n_{eff} = \frac{2\pi \cdot n_{eff}}{\lambda} \tag{2.31}$$

Here β is the propagation constant. So

$$\theta = \frac{\omega L}{c} = \frac{kc_o L}{c} = k \cdot n_{eff} \cdot 2\pi r = \frac{2\pi \cdot n_{eff} \cdot 2\pi r}{\lambda} = 4\pi^2 n_{eff} \frac{r}{\lambda} \tag{2.32}$$

According to Eqs. 2.28 and 2.30, we can get

$$E_{t1} = \frac{-\alpha + t \cdot e^{-j\theta}}{-\alpha t^* + e^{-\theta}} \tag{2.33}$$

$$E_{i2} = \frac{-\alpha \kappa^*}{-\alpha t^* + e^{-\theta}} \tag{2.34}$$

$$E_{t2} = \frac{-\alpha^*}{1 - \alpha t^* e^{j\theta}} \tag{2.35}$$

So the transmission power P_{t1} at the output waveguide can be expressed as

$$P_{t1} = |E_{t1}|^2 = \frac{\alpha^2 + |t|^2 - 2\alpha|t|\cos(\theta + \varphi_t)}{1 + \alpha^2|t|^2 - 2\alpha|t|\cos(\theta + \varphi_t)} \tag{2.36}$$

Here $t = |t|\exp(j\varphi_t)$, where $|t|$ is the coupling loss and φ_t is the phase delay.

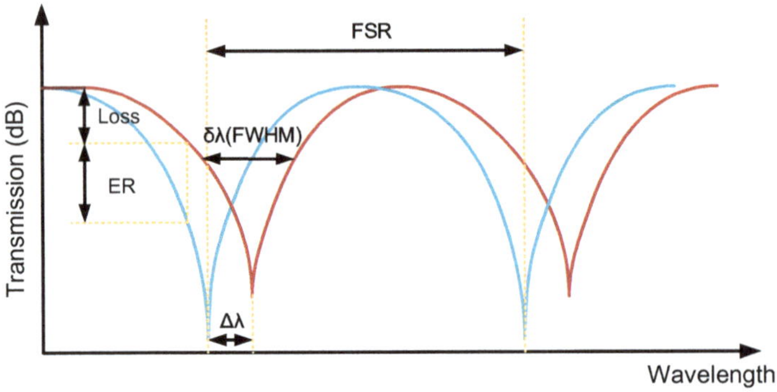

Fig. 2.9 Transmission characteristics of micro-ring modulator at different voltages

And the cycle power P_{i2} in the micro-ring can be expressed as

$$P_{t2} = |E_{t2}|^2 = \frac{\alpha^2(1-|t|^2)}{1+\alpha^2|t|^2 - 2\alpha|t|\cos(\theta+\varphi_t)} \tag{2.37}$$

At the resonance point

$$(\theta+\varphi_t) = 2\pi m \tag{2.38}$$

Here m is the integer. Equations 2.36 and 2.37 can be simplified as

$$P_{t1} = |E_{t1}|^2 = \frac{(\alpha-|t|)^2}{(1-\alpha|t|)^2} \tag{2.39}$$

$$P_{t2} = |E_{t2}|^2 = \frac{\alpha^2(1-|t|^2)}{(1-\alpha|t|)^2} \tag{2.40}$$

Especially, when the internal loss equals the coupling loss, i. e., $\alpha = |t|$, the transmission power is 0 according to Eq. 2.39. It is critical coupling.

Next, we further discuss some of the basic knowledge of MRR modulator, which is mainly referred to Ref. [73]. Figure 2.9 shows the transmission characteristics of MRR under different voltages. In the case of critical coupling, the resonance linewidth $\delta\lambda$ and quality factor Q can be expressed as

$$\delta\lambda \approx \frac{\lambda^2\alpha}{n_g\pi} \tag{2.41}$$

$$Q = \frac{\lambda}{\delta\lambda} \approx \frac{n_g\pi}{\lambda\alpha} \tag{2.42}$$

Here the group velocity n_g can be expressed as

$$n_g = n_{eff} - \lambda \frac{\partial n_{eff}}{\partial \lambda} \tag{2.43}$$

Here λ is the resonance wavelength. So the m-order resonance wavelength λ_m and (m+1)-order resonance wavelength λ_{m+1} can satisfy the relationship

$$m\lambda_m = n_m L \tag{2.44}$$

$$(m+1)\lambda_{m+1} = n_{m+1} L \tag{2.45}$$

Here n_m and n_{m+1} are the effective refractive index at the wavelength of λ_m and λ_{m+1}, respectively. If we let Eqs. 2.43–2.45, then we can get the free-spectrum range (FSR) as

$$FSR = \lambda_m - \lambda_{m+1} \approx \frac{\lambda^2}{n_g L} \tag{2.46}$$

Similarly, we also can get the resonant wavelength shift $\Delta\lambda$ caused by the effective refractive index change Δn_{eff}.

$$\Delta\lambda = \frac{\Delta n_{eff}}{n_g} \lambda \tag{2.47}$$

Photon in the micro-ring waveguide moves in group velocity and attenuates because of waveguide loss or coupling. The photon lifetime τ_{ph} is defined as the time when the intensity of light (the number of photons) attenuates to the original 1/e. Because the coupling loss in critical coupling is equal to the propagation loss, the photon needs to propagate in a distance of $1/(2\alpha)$ during a time of τ_p to get the attenuation of 1/e.

$$\tau_{ph} = \frac{n_g}{2c_o\alpha} \tag{2.48}$$

Therefore, the modulation bandwidth caused by photon lifetime limitation can be written as

$$f_{ph} = \frac{1}{2\pi\tau_{ph}} = \frac{c\alpha}{n_g\pi} \approx \frac{c}{\lambda Q} = \frac{c\delta\lambda}{\lambda^2} = \delta f \tag{2.49}$$

An important conclusion is that *modulation bandwidth does not exceed the resonant linewidth*. As a result, lower Q is easier to get higher bandwidth. In addition, the resistance and capacitance of the modulator can also affect the bandwidth performance. So the final modulator bandwidth is determined by both the lifetime of the photon and the resistance capacitance.

$$f_{3dB} = \frac{f_{ph} f_{RC}}{\sqrt{f_{ph}^2 + f_{RC}^2}} \tag{2.50}$$

2.4 Design and Optimization of Silicon Electro-Optic Modulator

The design of a silicon electro-optic modulator is different from that of a passive device. The latter only needs to be designed by using the general optical simulation software. Even some commercial simulation softwares provide some specific packages for some important passive devices. Users can only input design parameters to evaluate the performance of devices conveniently and accurately. The simulation of silicon electro-optic modulator is a complex opto-electronic hybrid simulation problem. It not only includes general optical mode analysis, but also semiconductor process simulation, RF electrode design and signal integrity analysis. There has not been another silicon photonic device that can be compared with in terms of the design parameters and the complexity of the design theory. And so far, there has not been a commercial software that provides a convenient and reliable design service for silicon-based modulators. For most researchers, the design optimization of the modulator is mainly based on the combination of commercial software and in-house programming. Therefore, the design theory of the modulator is systematically expounded, and the design process of the modulator is given, which are very important for realizing the engineering and standardization of silicon-based optical interconnections.

Figure 2.10 shows a schematic diagram of the design flow of a silicon electro-optic modulator. The whole design process is divided into three main steps. Starting from the design optimization of passive optical waveguides, the design and optimization of traveling-wave electrodes are finished after the design optimization of doping region. According to the actual design requirements, multiple rounds of design may be needed to converge to the optimal performance. Next, let's give a brief introduction to the content of each design step, and more details will be shown in the following section.

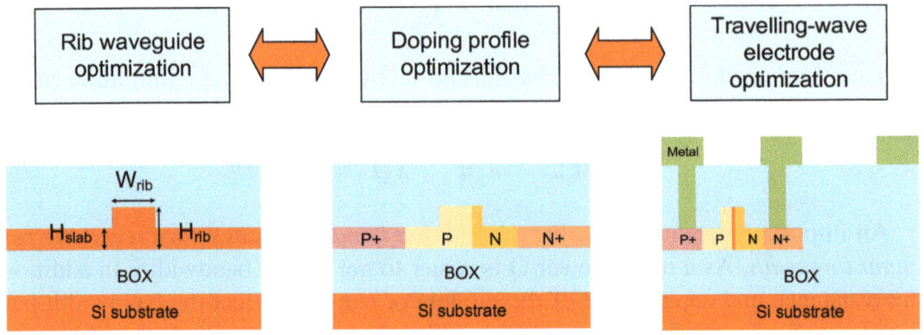

Fig. 2.10 Design process of silicon electro-optic modulator

1. Design and optimization of rib optical waveguide
   For silicon electro-optic modulator, the waveguide structure of phase shifter must be ridged because we need to connect the PN junction region with more mode field to the via and electrode on both sides. For the SOI wafer that we know the thickness of top layer and the buried oxide layer, the single-mode transmission condition is obtained by optimizing the slab thickness and the waveguide width. This part of the design can be accomplished by using the software such as FIMMWAVE, Rsoft, Comsol, or Lumerical MODE, which are commonly used for calculating the waveguide mode. When the size of a single-mode waveguide is obtained, the influence of the doping concentration on the refractive index of the material is not considered. Because the effect is very small, it does not break the single-mode transmission condition. However, a problem to consider is that the thickness of the slab layer will cause the change of capacitance, which will affect the bandwidth of the modulator. Therefore, we should not neglect this point in further optimization [74].
2. Design and optimization of doping region
   The design of the doped region directly affects the modulation efficiency and insertion loss of the modulator, so it needs to be considered carefully. On the one hand, the doping concentration of at least four doping regions needs to be chosen carefully. High doping concentration will improve the modulation efficiency, but increase the carrier absorption loss. The appropriate doping concentration needs to ensure good modulation efficiency and moderate insertion loss. At the same time, the selection of doping concentration also needs to communicate with the foundry to ensure that the design can be realized in the foundry. On the other hand, various PN junction structures and their specific sizes will also affect the performance of the modulator. For most of the basic lateral PN junction modulators, a key size is the location of PN interface. It is very sensitive to modulation efficiency and absorption loss. Because the doping concentration is not uniformly distributed in actual devices, a more accurate simulation method is used by semiconductor process simulation softwares, such as Silvaco TCAD software package. This software is used for carrier distribution simulation of the actual process conditions and the change of effective refractive index and absorption coefficient can be extracted. Finally the modulation efficiency and absorption loss can be calculated. Lumerical DEVICE also provides similar functions. Its advantages are simple and easy access, and it can simulate 3D interdigital and even more complex photonic crystals or polymer assisted complex modulator structures. And it is able to perform joint simulation with Lumerical FDTD and MODE to quickly calculate the information of modulation efficiency and absorption loss. However, it does not directly relate to the process parameters, so it is not a particularly accurate simulation method.

3. Design and optimization of travelling-wave electrode
   After determining the structure of the doping region, the last step is to optimize the traveling-wave electrode. In order to improve the transmission speed and bandwidth of the modulator, the ground-signal-ground (GSG) coplanar electrode structure is generally used. And in order to reduce the amplitude of the driving voltage, the electrode structure such as GSGSG can be used to carry out push-pull operation. The design of traveling-wave electrode is mainly based on the theory of microwave transmission line. The MZI modulator can be considered as a section of microwave transmission line. The different electrode materials and sizes will affect the characteristic impedance, the microwave attenuation coefficient and the speed of the microwave signal. The purpose of travelling-wave electrode optimization is to achieve impedance matching, speed matching and reduction of microwave attenuation. This part can be modeled and analyzed by using the numerical simulation software such as HFSS. In the simulation, the depletion layer of PN junction can be equivalent to a flat capacitor. The advantage of numerical simulation is not only to model the phase shifter of the modulator accurately, but also to analyze the influence of the detailed dimensions on the performance of the transmission line, and also to design and optimize the bending electrode between the pad and the active area. The disadvantage is that the speed is slow. If we want to establish the relationship between the design parameters and the performance of the device, it requires multiple scanning simulation. Another optimization method is the fully-analytical equivalent circuit model analysis. It can convert all the important parts of the active region into a series of capacitors and resistors, and these capacitors and resistors can be directly expressed by the analytical expression of the device size. The advantage of this method is that it can build a very intuitive physical model to deepen the designer's understanding on the device physical mechanism. The disadvantage is that some subtle places can not be accurately calculated, and some parasitic effects have to be ignored.

2.4.1 Numerical Simulation of Doping Region in a Lateral PN Junction Modulator

The key part of the phase shifter is PN junction. When the external driving electric signal is applied to the PN junction, the carrier concentration will change. According to the plasma dispersion effect, the change of the carrier concentration will lead to the change of the refractive index and absorption coefficient. Because these changes are different at various positions, it is difficult to estimate with a simple expression. A more accurate method is to overlap the distribution of the mode field of the waveguide with the change of the refractive index Δn(x,y) or the change of the absorption coefficient $\Delta\alpha(x, y)$, as shown in Eqs. 2.51 and 2.52. The overlap integral also represents the change of the equivalent refractive index of the whole waveguide mode field Δn_{eff} and the change of the absorption coefficient $\Delta\alpha$.

$$\Delta n_{eff} = \frac{1}{n_{eff}^0} \frac{\iint \Delta n(x, y) \cdot n^0(x, y) \cdot \left|E^0(x, y)\right|^2 dxdy}{\iint \left|E^0(x, y)\right|^2 dxdy} \tag{2.51}$$

$$\Delta \alpha = \frac{1}{\alpha^0} \frac{\iint \Delta \alpha(x, y) \cdot \alpha^0(x, y) \cdot \left|E^0(x, y)\right|^2 dxdy}{\iint \left|E^0(x, y)\right|^2 dxdy} \tag{2.52}$$

Here $E^0(x, y)$, $n^0(x, y)$, n_{eff}^0 and α^0 are the electric field component of the waveguide mode, the refractive index distribution of the material, the effective refractive index and the absorption coefficient of the waveguide mode, respectively.

According to the foregoing, the carrier concentration and doping concentration are closely related in each region. The design optimization of doping area is also influenced by process parameters besides structural parameters. Therefore, in order to accurately describe the doping concentration of the waveguide cross-section and its dynamic variation under different bias conditions, a semiconductor process simulation software must be used. Here, we have used a Silvaco TCAD package. First, the Athena is used to model the simulation region, and the initial doping concentration distribution of the cross-section of the whole waveguide is obtained. Secondly, the electrode is fed by Atlas, and the electrical properties of the structure are obtained, including the carrier concentration, which is derived as a separate data file. Finally, we use the MATLAB programming to collect these data files and convert it into a waveguide model. The influence of the carrier concentration on the refractive index and the absorption coefficient in different positions will be shown in the discrete meshes. In this way, after calculating the optical mode of the waveguide model, we can get the information of the distribution of the mode field and the overlap integral. Finally, the modulation efficiency $V_\pi L_\pi(V)$ and optical loss Loss(V) of the modulator under different bias V conditions can be calculated according to Eqs. 2.53 and 2.54.

$$V_\pi L_\pi(V) = \frac{\lambda}{2\Delta n_{eff}(V)} \times V \tag{2.53}$$

$$Loss(V) = 10 \log_{10}[\frac{-4\pi}{\lambda} \times \Delta\alpha(V)] \tag{2.54}$$

Silicon modulator has many kinds of PN junction structures, in which the lateral PN junction structure is the most widely concerned because of its simple process and easiness to realize high-speed modulation. A silicon modulator with lateral PN junction usually uses reverse bias to achieve carrier extraction, which results in the change of the depletion width of the PN junction, so such modulators are often referred to as carrier depletion modulators. Figure 2.11 shows a schematic of the cross-section of the active region of a lateral PN junction modulator for Silvaco Athena modeling. Although the design and processing of the modulator is based on the SOI wafer with a top thickness of 220 nm, we adopted 200 nm thickness in the actual simulation considering the loss of the top silicon layer during the process of oxidation. The slab thickness of the ridge waveguide is 100 nm and the width

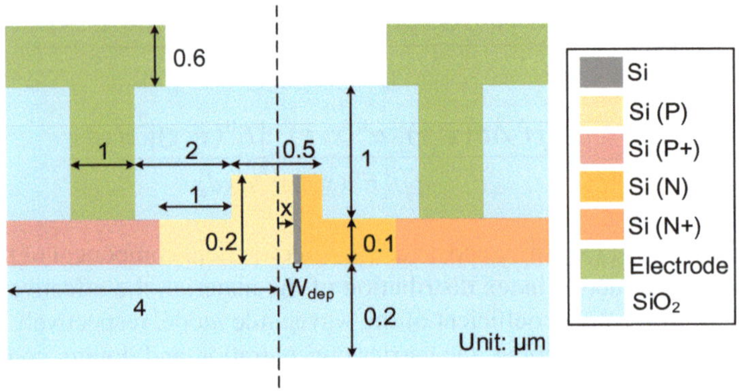

Fig. 2.11 Schematic diagram for modeling active region in a lateral PN-junction modulator

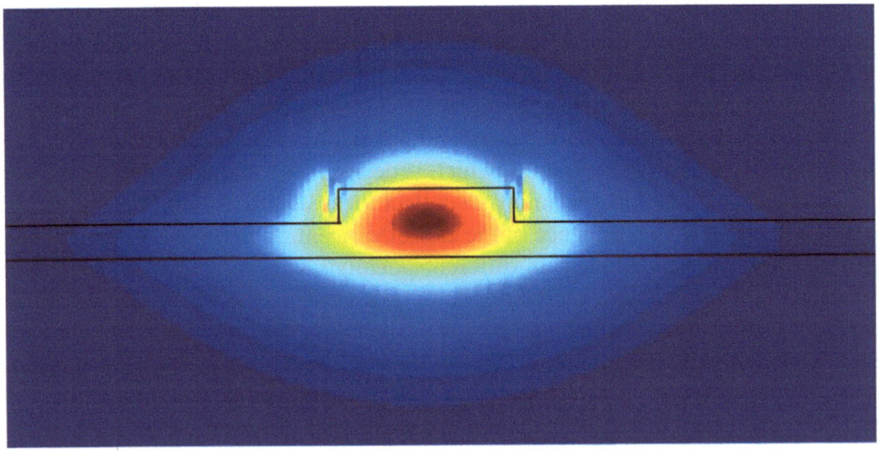

Fig. 2.12 Field distribution of TE mode in the rib doping waveguide. [2013] IEEE. Reprinted, with permission, from Ref. [75]

of the ridge waveguide is 500 nm, which can guarantee the a low-loss single-mode transmission. P-type heavy doping, P-type doping, N-type doping and N-type heavy doping area are realized by multiple ion implantation process from left to right, respectively. The heavy doping area is designed to ensure good ohmic contact with the electrode. A normal doping concentration in the waveguide with more mode field can ensure good modulation efficiency and appropriate absorption loss. Considering the actual technological conditions in a CMOS fabrication foundry, the target doping concentration in the heavily doped region is $1 \times 10^{20}/\text{cm}^3$ and the target doping concentration of PN junction is $3 \times 10^{17}/\text{cm}^3$–$5 \times 10^{17}/\text{cm}^3$.

For such a lateral PN junction structure, we used the program written by ourselves to analyze the mode in the doping rib waveguide with the above size. The distribution of its transverse electric (TE) mode is shown in Fig. 2.12.

After determining the above parameters, we need to optimize the other two important parameters. One is the initial depletion region thickness W_{dep}, and the other is the distance x from the middle line of the ridge waveguide to the left position of the initial depletion layer. When x is positive, the initial depletion layer is on the right side of the middle line; when x is negative, the initial depletion layer is on the left of the middle line. In Silvaco Athena, we divide the model into inhomogeneous meshes and carefully calculate the convergence. Figure 2.13 shows the relationship between the modulation efficiency $V_\pi L_\pi$ and the position x when PN junction is applied by –0.04 V bias voltage under different initial depletion layer width W_{dep}. For the case that the initial depletion layer width is smaller than W_{dep}, such as 0, 25 and 50 nm, we can find that the best modulation efficiency is achieved when x is located at 80–100 nm. This is because, according to the relationship of plasma dispersion effect, Eq. 2.16, under the same bias voltage, the effect of P doping carrier concentration on refractive index change is larger. It shows that the modulation efficiency is lower than $V_\pi L_\pi$. When x is smaller or larger, it is far away from the strongest interaction between the optical mode field and the carrier, so the modulation efficiency $V_\pi L_\pi$ will increase. When the initial depletion region width W_{dep} is large, such as 100 nm, the extraction of the carrier is weak because of the thick initial depletion layer. That's why the modulation efficiency remained relatively stable when x changes from –20 nm to 80 nm. When x > 100 nm, P doping region has filled most of the rib waveguide region and the position where depletion layer width changes is away from that where the optical mode field is larger, so the modulation efficiency of $V_\pi L_\pi$ will become larger.

Figure 2.13 shows the relationship between the absorption loss and the position x when the –0.04 V bias is applied to PN junction with different initial depletion layer thickness (W_{dep}). On the one hand, when x is fixed, the larger the initial depletion layer W_{dep}, the smaller the absorption loss. The main reason is that the existence of the depletion layer reduces the absorption of carriers. On the other hand, when the initial depletion layer thickness W_{dep} is fixed, the greater the position x, the

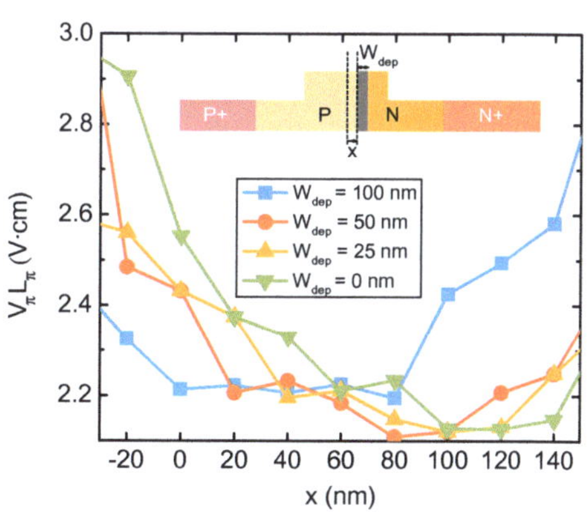

Fig. 2.13 Modulation efficiency as a function of offset x to the right of the middle of the rib waveguide with different depletion width W_{dep} at −0.04 V. [2013] IEEE. Reprinted, with permission, from Ref. [75]

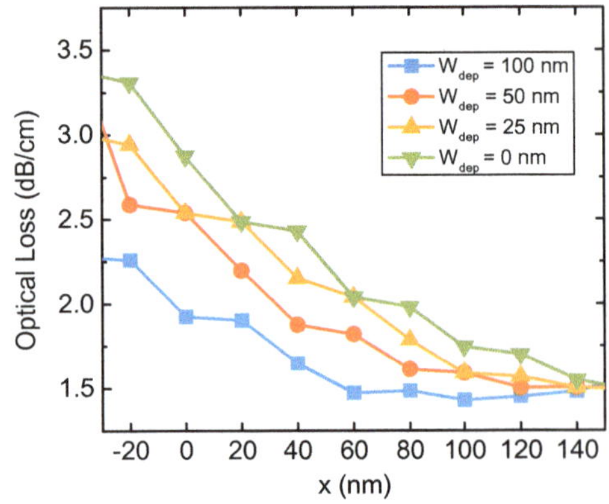

Fig. 2.14 Absorption loss as a function of offset x to the right of the middle of the rib waveguide with different depletion width W_{dep} at −0.04 V. [2013] IEEE. Reprinted, with permission, from Ref. [75]

smaller the absorption loss. This can also be explained by the plasma dispersion effect. According to Eq. 2.17, the absorption coefficient of P-type doping is smaller (Fig. 2.14).

Combined with the above simulation results, in subsequent processing, we selected a series of x between 60 and 100 nm, and a small number of W_{dep} for lateral PN modulators. Here, we choose a PN junction structure that achieves the best modulation efficiency of $V_\pi L_\pi$, namely x = 95 nm and W_{dep} = 0 nm, to investigate the effect of driving voltage on modulation efficiency $V_\pi L_\pi$ and absorption loss. When the reverse bias changes from 0.04 to 8 V, the simulation results are shown in Fig. 2.15. With the increase of the reverse bias voltage, the modulation efficiency becomes worse, and the performance of $V_\pi L_\pi$ increases gradually, while the optical absorption loss is obviously improved. The variation of modulation efficiency can be explained as the change of the width of the depletion layer is directly proportional to the square root of the reverse bias, while the decrease of the optical absorption loss

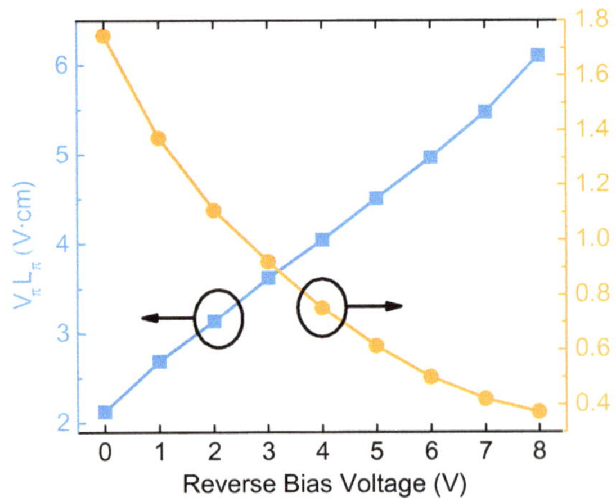

Fig. 2.15 Modulation efficiency $V_\pi L_\pi$ and absorption loss as a function of driving voltage when depletion width $W_{dep} = 0$ and position x = 0.095 μm. [2013] IEEE. Reprinted, with permission, from Ref. [75]

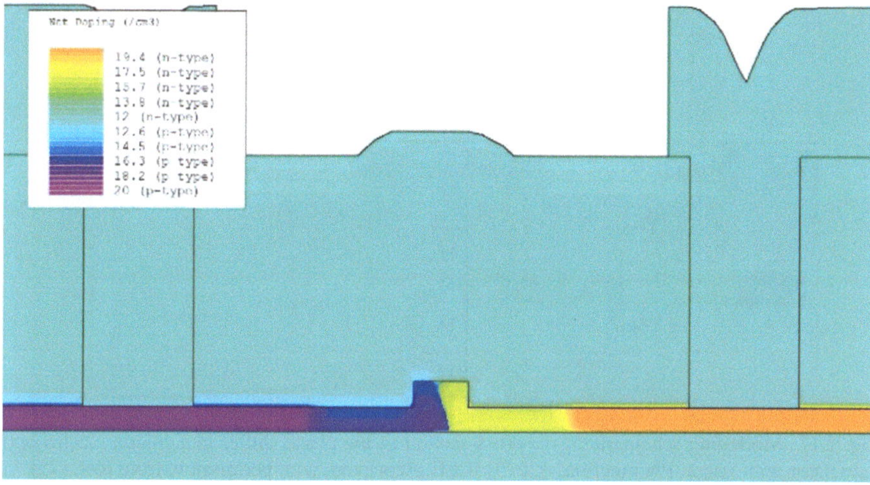

Fig. 2.16 Practical dopant distribution simulated in the software package Athena when initial depletion width $W_{dep} = 0$ and position $x = 0.095\,\mu m$. [2013] IEEE. Reprinted, with permission, from Ref. [75]

can be understood as the increase of the thickness of the depletion layer, resulting in the decrease of carrier concentration and the decrease of the light absorption. In addition, this trend also explains the necessity of low driving voltage of the modulator. This is not only because the realization of high power RF amplifier is difficult, and a low driving voltage can ensure the modulation efficiency good enough, but also it can reduce the power consumption in silicon-based optical interconnections.

Figure 2.16 shows the doping simulation results when the initial depletion region width is $W_{dep} = 0$ and the position $x = 0.095\mu m$. It is obvious that the PN junction interface is not vertical, and the carrier concentration in the same doping region is also related to the position. All of these illustrate the necessity of using semiconductor process simulation software to optimize the silicon electro-optic modulator, which is unable to be replaced by other methods.

2.4.2 *Fully-Analytical Equivalent Model Simulation of Silicon MZI Electro-Optic Modulator*

In order to optimize the high-frequency characteristics of silicon electro-optic modulator, HFSS and other high-frequency numerical simulation software will be used in general. Its advantage is that it can make a complete modeling including most details of the modulator, so as to ensure that the simulation results are accurate enough. But the disadvantage is that it needs a longer simulation time, and the rela-

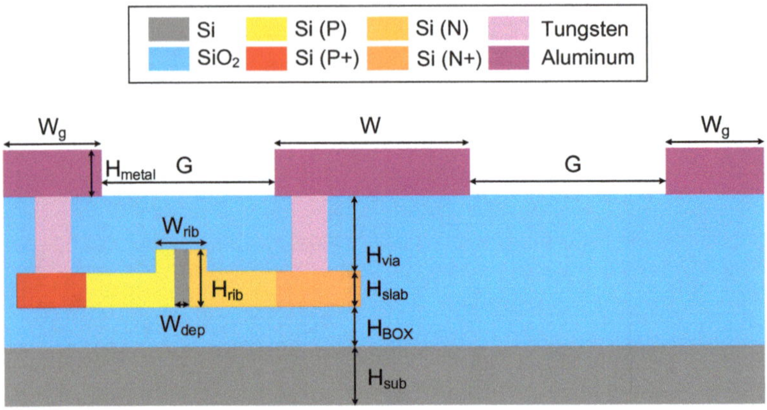

Fig. 2.17 Modelling schematic of the cross-section of the phase shifter in a silicon electro-optic modulator with lateral PN junction. [2013] IEEE. Reprinted, with permission, from Ref. [75]

tionship between device size and performance is not clear enough. Usually, a lot of sweeping simulation is needed to find out the design rule. For the traditional lithium niobate modulator, the fully-analytical equivalent circuit model of the microwave transmission line has been established. However, when the carrier-depletion silicon electro-optic modulator with lateral PN junction is considered, the situation becomes more complicated. This is because the structure of the silicon modulator is not simply homogeneous dielectric layer plus the structure of the top electrode, and the doping concentration of the waveguide area is also different. All of these challenge the establishment of a precise equivalent circuit model. Therefore, few studies about the equivalent model of silicon electro-optic modulator has been done in the past few years. Some of these work are based on fitting the equivalent circuit model from numerical simulation results or experimental data. Until 2012, Dr. Yu Hui from Ghent university first proposed a forward equivalent circuit model for silicon carrier-depletion electro-optic modulator with lateral PN junction [76]. And our work in this section is partially based on his model.

Figure 2.17 shows a schematic of a complete cross-section of the phase shifter of a silicon electro-optic modulator with lateral PN junction. Unlike previous models used in semiconductor process simulation softwares (Fig. 2.11), when considering the high-frequency characteristics of the modulator, we must take into account the influence of electrodes. Here, we use the GSG CPW transmission line for the phase shifter. The model parameters and the constants used in the equivalent circuit model of the transmission line are summarized in Table 2.1.

In the microwave frequency, the equivalent model of transmission line of the modulator supports three kinds of microwave transmission mechanisms. These microwave transmission mechanisms have been widely studied in other people's work. Here we briefly summarize here:

Table 2.1 Design parameters for the fully-analytical simulation of MZI electro-optic modulator

Parameter	Definition	Value	Unit
W	Width of signal electrode	4–16	μm
W_g	Width of ground electrode	5W	μm
G	Gap between signal and ground electrodes	2–8	μm
H_{metal}	Electrode height	0.5	μm
W_{dep}	Width of depletion layer	0.07–0.15	μm
N_{Si}	Doping concentration of PN junction	$0.5\times10^{17}-5\times10^{17}$	$/cm^3$
ρ_{Si}	Resistivity of PN junction	Lookup according to N_{Si}	$\Omega\cdot$ cm
σ_{Si}	Conductivity of PN junction	$1/\rho_{Si}$	S/m
W_{rib}	Width of rib waveguide	0.5	μm
H_{rib}	Height of rib waveguide	0.2	μm
H_{slab}	Slab height	0.1	μm
H_{via}	Via height	1.4	μm
H_{BOX}	Height of Buried oxide layer	2	μm
H_{sub}	Height of silicon substrate	500	μm
ρ_{sub}	Resistivity of silicon substrate	0.1	$\Omega\cdot$ cm
n_o	Group index of optical mode	3.64	N.A.
Z_s	Source impedance	50	Ω
Z_t	Terminator impedance	50	Ω
l	Transmission length	0.001	m
v_g	Amplitude of driving signal	1	V
f_m	Frequency of driving signal	1–100	GHz
ω_m	Angular frequency of driving signal	$2\pi f_m$	rad/s
ω_0	Minimum angular frequency of driving signal	$2\pi\times10$ MHz	rad/s
ϵ_0	Permittivity of vacuum	8.85×10^{-12}	F/m
ϵ_{Si}	Relative permittivity of silicon	11.9	N.A.
ϵ_{SiO2}	Relative permittivity of SiO_2	1.56	N.A.
c	Velocity of light	3×10^8	m/s
μ_0	Permeability of vacuum	$4\pi\times10^{-7}$	H/m
κ_{Al}	Conductivity of aluminum electrode	3.8×10^7	S/m

1. If the P-doping and N-doping regions in the space-charge region are lightly doped, they will show a lower dielectric loss tangent at the microwave frequency. At this time, silicon in the P doped region and the N doped region shows the properties of the medium, and the TE and the transverse magnetic (TM) fields are free to penetrate them. Such a transmission mode is called the "*quasi-TEM mode in medium* ".
2. If the P doping region and the N doped region are moderately doped, the dielectric relaxation frequency will be higher than the operation frequency at this time. The external carriers in the space-charge zone can react quickly to the AC electric

field, thus avoiding the interior of the heavily P-doped and N-doped regions. Therefore, the electric field of the microwave driving signal is confined to the space-charge area. At the same time, if the widths of the P doped and N doped regions are less than the skin depth, the TM mode field can penetrate them while TE is not. The separation of the two energy in space leads to a "*slow-wave mode*". Its equivalent permittivity is larger than the dielectric constant of the other media in the transmission line.

3. If the P-doping and N-doping regions are heavily doped, even if their width is larger than the skin depth, the two regions will be similar to the damaged semiconductor walls, preventing the electric field and magnetic field from penetrating them. This is called "*skin effect mode*".

In a general silicon optical modulator, there is no heavily doping region. Therefore, only the first two modes can be considered, namely, the dielectric quasi TEM mode and the slow-wave mode. Furthermore, previous researchers have found that the quasi TEM mode analysis can be used to deal with the slow-wave mode of the CPW electrode in the micron scale. Therefore, all the structures in the modulator can be processed by the quasi TEM mode analysis method, and then the equivalent circuit model can be obtained.

In addition, we used partial capacitance technology to extract the equivalent circuit model as shown in Fig. 2.17. This method assumes that all the media interfaces are in the direction of the electric field line, so the magnetic wall will appear at the interface. Based on this assumption, the CPW electrode on multilayer substrate can be decomposed into several CPW electrodes on different single-layer substrates, which have corrected dielectric constant. The capacitance of each CPW electrode per unit length can be calculated by coplanar mapping method, so the actual CPW capacitance is the sum of these partial capacitance. Although partial capacitance technology is an approximation, it has been proved to be effective for the problem of CPW with multi-layer rectangular substrates. This method can be used in the modulator structure because the electric field in the carrier-depletion region is parallel to that in the Si waveguide and buried oxide layer.

Based on the partial capacitance technology and the coplanar mapping method, we extracted electrical model in each region of Fig. 2.17 as shown in Fig. 2.18a, and summarize it as the equivalent model of the microwave transmission line, which is shown in Fig. 2.18b. Next, we will discuss how to extract the capacitance and resistance in the equivalent circuit model and its calculation method.

The capacitance of air above the device C_{air} can be expressed as

$$C_{\text{air}} = 4\epsilon_0 \frac{K(k_0')}{K(k_0)} - 2\epsilon_0 \frac{K(k_1')}{K(k_1)} \tag{2.55}$$

Here $K(k_i)$ is the complete elliptic integral of the first kind, which can be solved by the ellipke function of MATLAB. And k_i can be expressed by Eqs. 2.56 and 2.57.

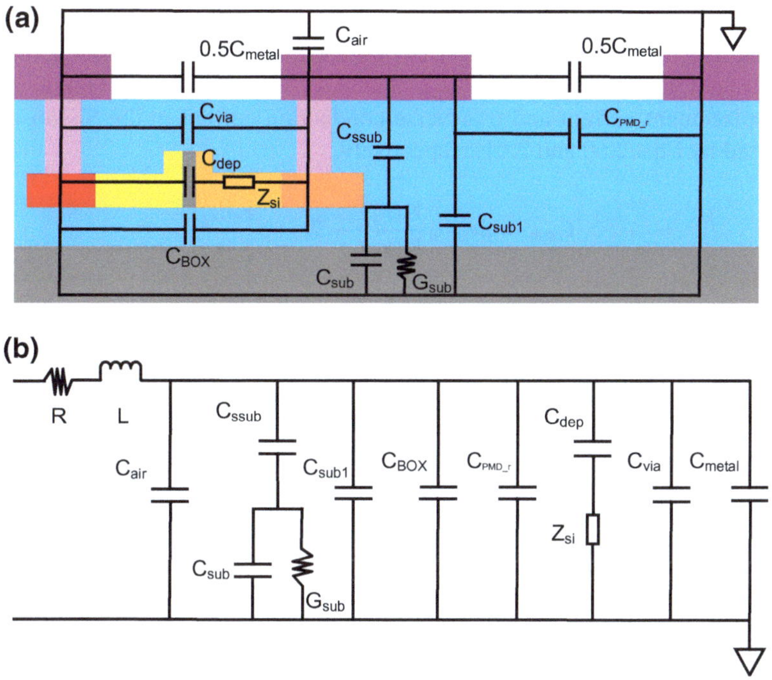

Fig. 2.18 Equivalent circuit model of a silicon optical modulator with lateral PN junction. [2013] IEEE. Reprinted, with permission, from Ref. [75]

$$k_0 = \frac{x_c}{x_b}\sqrt{\frac{x_b^2 - x_a^2}{x_c^2 - x_a^2}} \tag{2.56}$$

$$k_i = \frac{\sinh(\frac{\pi x_c}{2h_i})}{\sinh(\frac{\pi x_b}{2h_i})}\sqrt{\frac{\sinh^2(\frac{\pi x_b}{2h_i}) - \sinh^2(\frac{\pi x_a}{2h_i})}{\sinh^2(\frac{\pi x_c}{2h_i}) - \sinh^2(\frac{\pi x_a}{2h_i})}} \tag{2.57}$$

Here $k_i' = \sqrt{1 - k_i^2}$ (i = 1, 2, 3, 4), $h_1 = H_{via}+H_{slab}+H_{BOX}+H_{sub}$, $h_2 = h_1$-H_{sub}, $h_3 = h_2$-H_{BOX}, $h_4 = h_3$-H_{slab}, $x_a = W/2$, $x_b = x_a+G$, and $x_c = x_b+W_g$.

Similarly, the capacitance C_{BOX} of the buried oxide region and the capacitance C_{PMD_r} of the right upper cladding can also be solved by Eqs. 2.58 and 2.59.

$$C_{BOX} = 2\epsilon_0\epsilon_{SiO2}\frac{K(k_2')}{K(k_2)} - 2\epsilon_0\epsilon_{SiO2}\frac{K(k_3')}{K(k_3)} \tag{2.58}$$

$$C_{PMD\_r} = \epsilon_0\epsilon_{SiO2}\frac{K(k_3')}{K(k_3)} \tag{2.59}$$

The Si substrate, which is a layer of semiconductor with non-zero conductivity, is different from the buried oxide layer. So we need to use four electrical components to describe it, namely C_{sub}, G_{sub}, C_{ssub}, and C_{sub1}. We use C_{sub} and G_{sub} to describe the displacement current and transverse conduction current in the Si substrate, as expressed by Eqs. 2.60 and 2.61, respectively.

$$C_{sub} = 2\epsilon_0\epsilon_{Si}\frac{K(k_1')}{K(k_1)} - 2\epsilon_0\epsilon_{Si}\frac{K(k_2')}{K(k_2)} \tag{2.60}$$

$$G_{sub} = \frac{2}{\rho_{sub}}\left[\frac{K(k_1')}{K(k_1)} - \frac{K(k_2')}{K(k_2)}\right] \tag{2.61}$$

C_{ssub}, $C_{sub}$and G_{sub} are connected in series to represent the capacitance between the signal electrode and Si substrate. It can be expressed by Eq. 2.62.

$$C_{ssub} = \epsilon_0\epsilon_{SiO2} \times \frac{W}{H_{via} + H_{slab} + H_{BOX}} \tag{2.62}$$

Here, because the Si waveguide is very thin, we ignore the effect on this part of the capacitance. The capacitance C_{sub1} is not actually existing, but it can keep the model stable in the high frequency range. When the frequency is high enough, the conduction current is negligible compared to the displacement current, so the Si substrate can be regarded as a media layer. According to the partial capacitance technology, the dielectric layer should be equivalent to a capacitor C_{sub}, while the parallel capacitor C_{sub1} can guarantee the total capacitance converging to C_{sub} at the high frequency range. So C_{sub1} can be expressed by Eq. 2.63.

$$C_{sub1} = C_{sub} - \frac{C_{sub}C_{ssub}}{C_{sub} + C_{ssub}} \tag{2.63}$$

In order to describe Si waveguide and its upper cladding, we have treated it as the impedance Z_{Si} of doped Si, the PN junction capacitance C_{dep} and the capacitance C_{via} between two vias. Because the Si waveguide is very thin, the fringe field can not be ignored when calculating the PN junction capacitance C_{dep}. If the influence of ridge type is ignored, the doping Si in the middle of the depletion region can be equivalent to two plates with finite length, so the reverse biased PN junction can be equivalent to a parallel plate capacitor. Dr. Yu gave the revised formula of the PN junction capacitance C_{dep} as follows:

$$C_{dep} = \left(C_1 - \epsilon_0\epsilon_{SiO2}\frac{H_{rib}}{W_{dep}}\right)\frac{\epsilon_{SiO2}}{\epsilon_{Si}} + \epsilon_0\epsilon_{Si}\frac{H_{rib}}{W_{dep}} \tag{2.64}$$

Here C_1 can be referred to Ref. [77], which is no longer described here. The first part of Eq. 2.64 takes into account the influence of the fringe field, while the second part represents the field within the Si.

In addition, the impedance Z_{Si} of doped Si, the capacitance C_{via} between two vias, and the capacitance C_{metal} between the signal and grounding electrodes are solved by Eqs. 2.65, 2.66 and 2.67.

$$Z_{Si} = \frac{\rho_{Si}}{1 + j\omega_m \rho_{Si} \epsilon_0 \epsilon_{Si}} \left(\frac{W_{rib} - W_{dep}}{H_{rib}} + \frac{G - W_{rib}}{H_{slab}} \right) \tag{2.65}$$

$$C_{via} = \epsilon_0 \epsilon_{Si} \frac{H_{via}}{G} \tag{2.66}$$

$$C_{metal} = 2\epsilon_0 \epsilon_{Si} \frac{H_{metal}}{G} \tag{2.67}$$

There are still three parts need to be considered in the circuit model.

- The line resistance R and the line conductance L represent the longitudinal current in the electrode. At low frequency, the line resistance R can be calculated directly by Ohm law, and the circuit conductance L can also be solved from the static magnetic field theory. These two parameters are constants in this frequency range. However, as the frequency increases, they become frequency dependent because of the skin effect of imperfect metals. In the skin-effect area, the conductance is divided into two parts, namely the internal conductance and the external conductance. The external conductance is frequency dependent and can be extracted from the CPW capacitance of the air with the same size. The internal conductance is calculated by the Wheeler's rule [78]. The calculation of the circuit resistance R is also based on the same theorem. Heinrich et. al., gave a closed calculation formula for R and L in Ref. [78], which is no longer described here.
- The resistance R_{Si} represents the longitudinal current in the Si substrate and Si waveguide layer. This is due to the interaction of Si with the magnetic field. This longitudinal current is parallel to the internal current of the metal, and it also causes loss. However, because of the high resistivity of the thin Si waveguide layer and the actual SOI substrate, R_{Si} is higher than the impedance given by R and L, and we can ignore it.

In summary, we have obtained the closed analytical formula of the various electrical components in the equivalent circuit model of the modulator. Furthermore, the parallel admittance Y and the series impedance Z of the microwave transmission line per unit length can be obtained, which are expressed by Eqs. 2.68 and 2.69 respectively.

$$\begin{aligned}
&\mathrm{Y} = \mathrm{j}\omega_{\mathrm{m}}\mathrm{C_t} + \mathrm{Y_1} + \mathrm{Y_2} \\
&\mathrm{C_t} = \mathrm{C_{air}} + \mathrm{C_{sub1}} + \mathrm{C_{BOX}} + \mathrm{C_{PMD_r}} + \mathrm{C_{via}} + \mathrm{C_{metal}} \\
&\mathrm{Y_1} = \cfrac{1}{\cfrac{1}{\mathrm{j}\omega_{\mathrm{m}}\mathrm{C_{sub}} + \mathrm{G_{sub}}} + \cfrac{1}{\mathrm{j}\omega_{\mathrm{m}}\mathrm{C_{ssub}}}} \\
&\mathrm{Y_2} = \cfrac{1}{\cfrac{1}{\mathrm{j}\omega_{\mathrm{m}}\mathrm{C_{dep}}} + \mathrm{Z_{Si}}}
\end{aligned} \tag{2.68}$$

$$Z = R + j\omega L \tag{2.69}$$

As a result, the propagation constant γ and the characteristic impedance Z_0 can be given by Eqs. 2.70 and 2.71, respectively.

$$\gamma = \sqrt{ZY} \tag{2.70}$$

$$Z_0 = \sqrt{\frac{Z}{Y}} \tag{2.71}$$

The effective dielectric constant ϵ_{eff} and the attenuation coefficient α of the transmission line are the real and imaginary parts of the propagation constant γ, respectively.

$$\epsilon_{\mathrm{eff}} = \mathrm{Re}(\gamma) \tag{2.72}$$

$$\alpha = Im(\gamma) \tag{2.73}$$

Although we have obtained some parameters of the transmission line, these parameters do not represent the performance of the modulator. The high-frequency characteristics of the modulator need to be obtained by solving the frequency response of the modulator. It is assumed that the angular frequency of the driving signal is ω_m, and the amplitude is V_g. When a photon passes through the phase shifter, the average voltage $V_{\mathrm{avg}}(\omega_{\mathrm{m}})$ between the signal electrode and the ground electrode can be expressed by Eqs. 2.74 and 2.79.

$$V_{\mathrm{avg}}(\omega_{\mathrm{m}}) = \frac{\mathrm{V_g}(1+\rho_1)\exp(\mathrm{i}\beta_o \mathrm{l})}{2[\exp(\gamma \mathrm{l}) + \rho_1\rho_2\exp(-\gamma \mathrm{l})]}(\mathrm{V_+} + \rho \mathrm{V_-}) \tag{2.74}$$

$$V_{\pm} = \exp(\pm \mathrm{i}\phi_{\pm})\frac{\sin\phi_{\pm}}{\phi_{\pm}} \tag{2.75}$$

$$\phi_{\pm} = \frac{(-i\gamma \mp \beta_o)l}{2} \tag{2.76}$$

$$\rho_1 = \frac{Z_0 - Z_S}{Z_0 + Z_S} \tag{2.77}$$

$$\rho_2 = \frac{Z_t - Z_0}{Z_0 + Z_t} \tag{2.78}$$

$$\beta_o = \frac{\omega_m}{c} n_o \tag{2.79}$$

Here Z_S and Z_t are the source impedance and terminal impedance of the transmission line, respectively. l is the length of the transmission line (also the length of the electrode and phase shifter), and n_o is the group refractive index of the optical mode field.

Under a small driving voltage, the modulation depth is directly proportional to the average driving voltage on the depletion region V_{dep}, while the latter can be multiplied by the ratio of $V_{avg}(\omega_m)$ and C_{dep} to total capacitance. When we normalize the modulation depth, we can get the frequency response of the modulator as follows:

$$m(\omega_m) = \left| \frac{V_{dep}(\omega_m)}{V_{dep}(\omega_0)} \right| = \left| \frac{(1 + j\omega_0 C_{dep} Z_{Si}) V_{avg}(\omega_m)}{(1 + j\omega_m C_{dep} Z_{Si}) V_{avg}(\omega_0)} \right| \tag{2.80}$$

Here ω_0 is the minimum output angle frequency of the microwave source, and it is 10 MHz here. Note that Eq. 2.80 has taken into account three elements in the design of microwave transmission lines, such as velocity matching, impedance matching, and microwave loss. And the 3 dB bandwidth is defined as the frequency where $m(\omega_m)$ decreased by 50%.

According to the formula above, we have written the MATLAB program. There are four important parameters to be discussed for optimizing the bandwidth, namely doping concentration N_{Si} of PN junction, depletion width W_{dep} of PN junction, signal electrode width W, gap G between signal electrode and grounding electrode. In the calculation, we get the resistivity of P- and N- based on the doping concentration N_{Si} of PN junction, and then put the average resistivity ρ_{Si} into the corresponding formula.

Figure 2.19 shows the relationship between the 3 dB bandwidth and the depletion width W_{dep} when the concentration $N_{Si} = 3 \times 10^{17}/cm^3$ of PN junction and the width $W = 6 \mu m$ of the signal electrode. Other parameters are shown in Table 2.1. With the increase of the depletion width W_{dep}, the 3 dB bandwidth will also increase. This is mainly due to the reduction of PN junction capacitance C_{dep}. Moreover, the smaller the gap G between the signal electrode and the ground electrode, the larger the series impedance. So the bandwidth will also increase. According to the transient PN junction model and the actual DC bias conditions, the depletion width $W_{dep} = 0.15 mm$ is possible. In the following simulation, we will use this W_{dep}.

In addition to the depletion width W_{dep}, the doping concentration N_{Si} of PN junction has a great effect on the bandwidth. Figure 2.20a–c show the relationship between the 3 dB bandwidth and the doping concentration N_{Si} of the PN junction

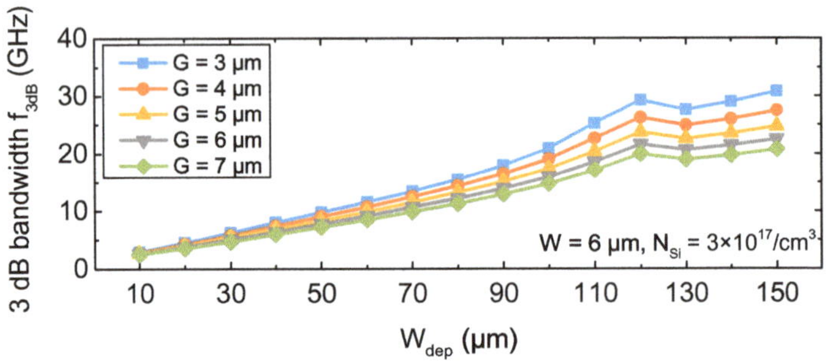

Fig. 2.19 3 dB EO bandwidth as a function of the width W_{dep} of the depletion region when doping concentration of PN junction $N_{Si} = 3 \times 10^{17}/cm^3$ and width of signal electrode $W = 6\mu m$

under different electrode sizes W and G. Because the higher the doping concentration, the lower the series impedance, a larger bandwidth can be achieved but this is at the expense of a larger carrier absorption loss. When the doping concentration of the PN junction $N_{Si} = 3 \times 10^{17}/cm^3$ and signal electrode width $W = 6\mu m$, the 3 dB bandwidth is more than 30 GHz for a electrode gap G <3 μm.

We further gave a contour map of the 3 dB bandwidth under different electrode sizes when the concentration of PN junction is $N_{Si} = 3 \times 10^{17}/cm^3$ and the depletion width is $W_{dep} = 0.15\ \mu m$. It is shown that when the gap between the electrodes is $G = 2\mu m$, the 3 dB bandwidth is up to 35 GHz. However, such a small electrode gap may make the microwave signal in the electrode affecting the mode propagation in the waveguide, causing additional reflection or loss. Therefore, we choose $W = 6\mu m$ and $G = 3\mu m$ as the size of the CPW electrode which can guarantee 3 dB bandwidth of 33 GHz and also support high-speed transmission over 40 Gb/s (Fig. 2.21).

Figure 2.22a–d show each important parameter of microwave transmission line after optimization, including attenuation coefficient α, microwave effective index n_m, real part Real(Z_0) and imaginary part Imag (Z_0) of characteristic impedance. The calculation of these parameters by the equivalent circuit model can be compared with the numerical simulation results by HFSS. The HFSS numerical simulation can not calculate the modulator performance such as frequency response or bandwidth, but it is also of great significance because these parameters are the intermediate products for evaluating the performance of the modulator,

The fully-analytical equivalent circuit model can also deepen the understanding of the modulator design theory. For example, in the general simulation, the impedance of the microwave terminator is $Z_t = 50\Omega$. Here, we have calculated the effect of different terminators Z_t on the frequency response of modulator under our optimized electrode sizes, as shown in Fig. 2.23. We can find that for small terminal impedance, it can improve the bandwidth performance of the modulator. It has been analysed that this may be because small impedance causes some reflection, resulting in the pre-strengthening of high-frequency microwave signals. But by using the analytical model, it can be found that the high frequency microwave signal is not strengthened,

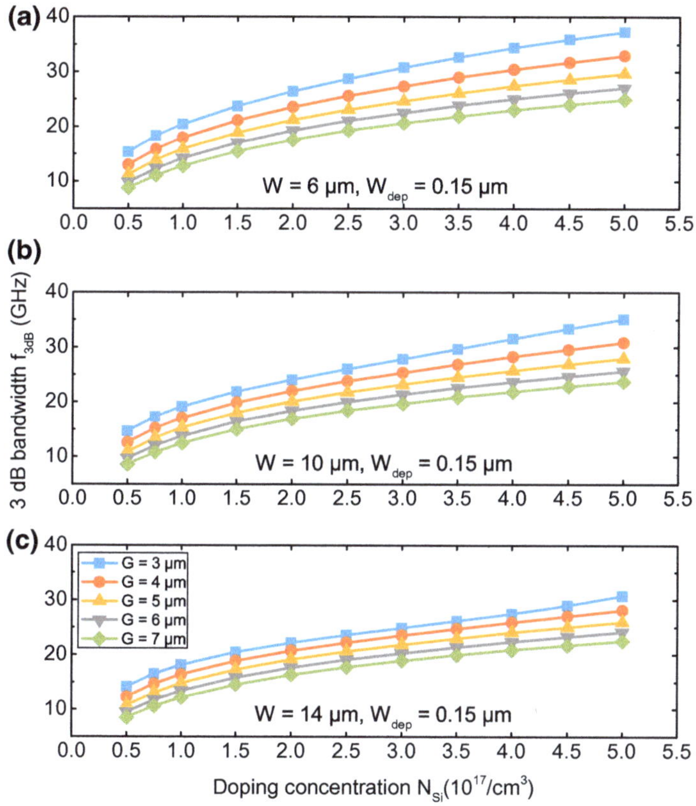

Fig. 2.20 3 dB EO bandwidth as a function of the doping concentration N_{Si} of the PN junction with different electrode sizes W and G. [2013] IEEE. Reprinted, with permission, from Ref. [75]

Fig. 2.21 Contour map of 3 dB modulation bandwidth with various W and G. [2013] IEEE. Reprinted, with permission, from Ref. [75]

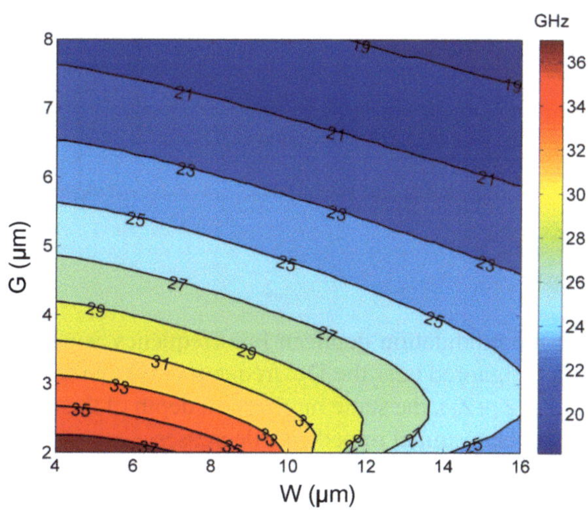

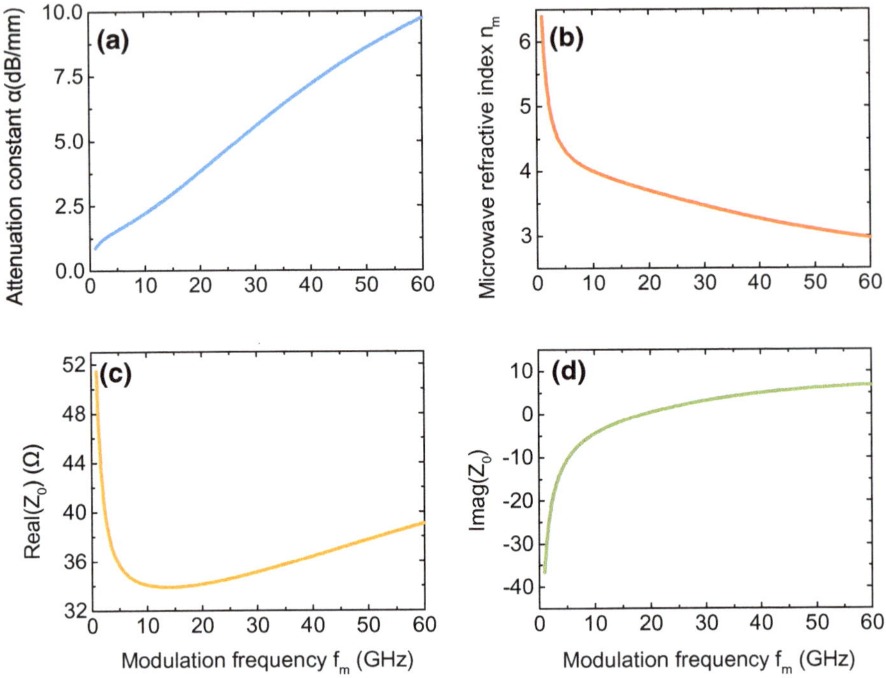

Fig. 2.22 Parameters of transmission line with optimized electrode size

Fig. 2.23 Effect of terminator impedance Z_t on modulator frequency response

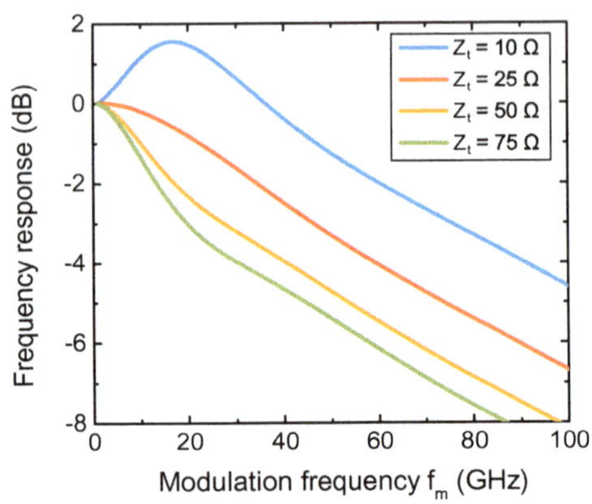

but the modulation depth in low frequency is suppressed. The resistance of CPW can be ignored near the DC frequency. When the voltage of the transmission line is $V_g Z_t/(Z_s+Z_t)$, the static modulation depth of the transmission line will also decrease when the terminal impedance decreases.

2.5 44 Gb/s Carrier Depletion Silicon Electro-Optic Modulator Based on MZI Structure

All of the silicon optical modulators we have designed were processed on the silicon photonic CMOS platform of Grace semiconductor manufacturing corporation in Shanghai. The platform is based on the 130 nm standard CMOS technology. The SOI wafer with a top silicon thickness of 220 nm was used. The thickness of the buried oxide layer is 2 μm, so as to prevent the optical mode from leaking to the substrate, resulting in additional transmission loss. At the same time, the minimum feature size of the waveguide is 150 nm and the minimum gap is 200 nm by using 248 nm UV photolithography. In order to achieve two kinds of waveguides, strip waveguide and ridge waveguide, a 220 nm top silicon etching is decomposed into two-step etching process. Meanwhile, in order to improve the sidewall roughness of the optical waveguide, the SiN layer is used as the etching mask. The processing steps are shown in Fig. 2.24 and described as follows:

1. A SiN layer is grown on the SOI wafer. The SiN layer is used as a mask for the subsequent silicon etching process. Because the stress coefficient between SiN layer and Si wafer does not match, in order to alleviate the stress between SiN and Si wafer, a thin layer SiO_2 is usually first grown on Si wafer as buffer layer before growing SiN layer.
2. The Si waveguide region is defined on the wafer by 248 nm deep ultraviolet photolithography (DUV).
3. The SiN mask is removed by inductively coupled plasma reactive ion etching (ICP-RIE), and the photoresist is used as an etching mask.
4. The ridge waveguide and the strip waveguide are realized by the second photolithography. The photoresist is covered by the SiN mask at the ridge waveguide, while the photoresist is removed by the developing method at the strip waveguide.
5. First Si etching.
6. The etching photoresist in the last etching is removed without photolithography.
7. Second Si etching. After the etching process, the strip waveguide and the ridge waveguide are realized simultaneously.
8. After removing the SiN mask layer by wet etching, the basic structure of the optical waveguide is realized.
9. The third photolithography realizes all the N doping regions. First ion implantation.
10. The fourth photolithography realizes all the P doping regions. Second ion implantation.
11. The fifth photolithography realizes the N+ doping area. Third ion implantation.
12. The sixth photolithography realizes the P+ doping area. Fourth ion implantation.
13. A SiO_2 protection layer is deposited.
14. The via structure is defined by the seventh photolithography. Interlevel dielectric deposition (ILD), deposition of W metal and chemical mechanical polishing (CMP).

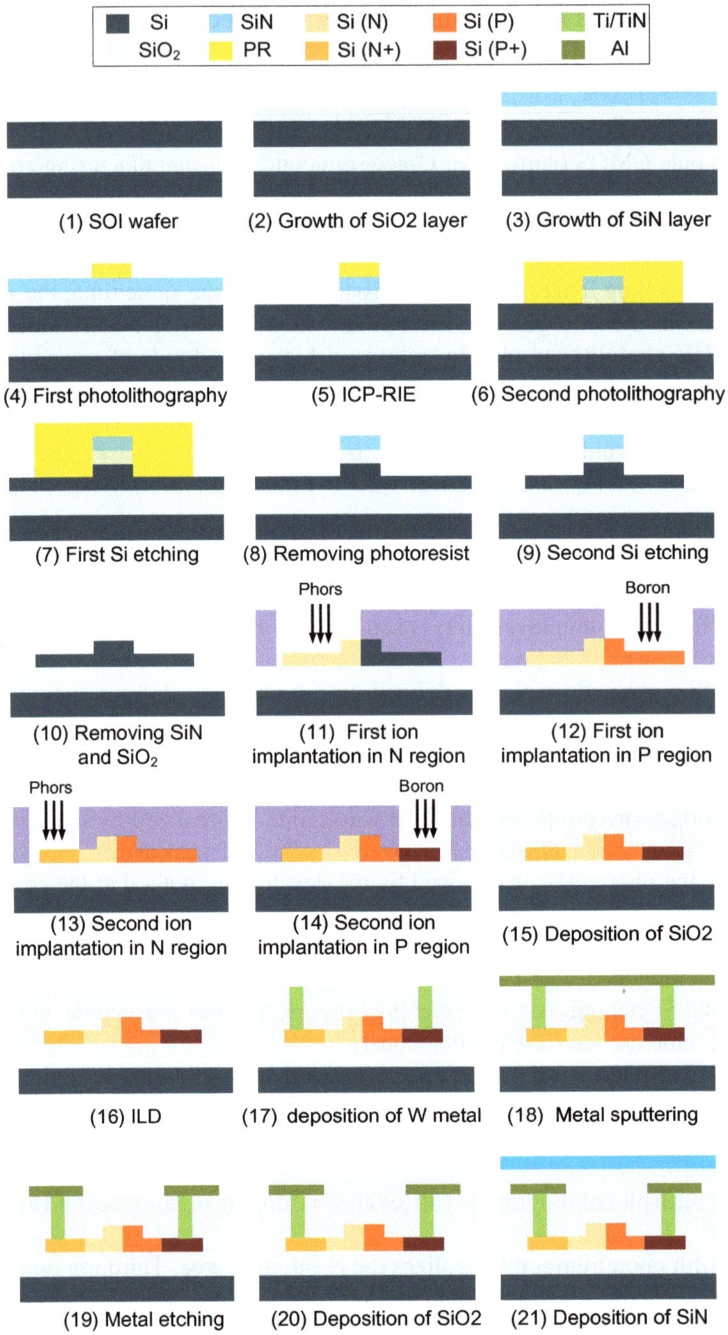

Fig. 2.24 Process flow for fabricating silicon optical modulators in 130 nm CMOS process

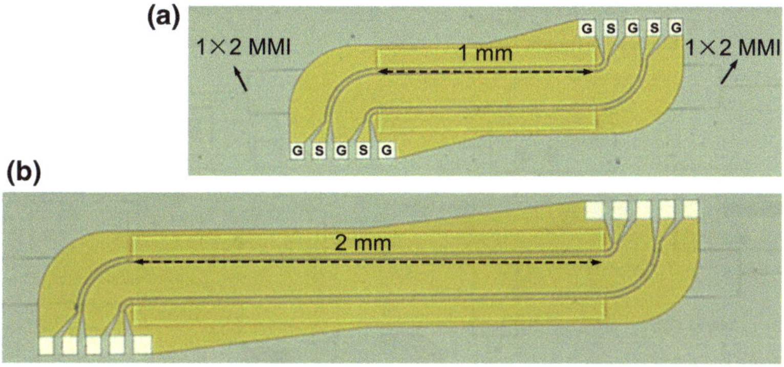

Fig. 2.25 Microscopes of the MZI modulator samples with **a** 1 mm and **b** 2 mm phase shifter lengths. [2013] IEEE. Reprinted, with permission, from Ref. [75]

15. Electrode metal sputtering.
16. The eighth photolithography defines the structure of the Al electrode.
17. Finally, a SiO_2 layer and a SiN layer are deposited to protect the device.
18. For the edge coupled devices, the edge of the chip needs to be deeply etched, and then the dicing is carried out.

Based on the above process, we have processed a series of modulator samples with different design parameters. Figure 2.25a–b show the microscopes of the MZI modulator samples with phase shifter lengths of 1 mm and 2 mm, respectively.

In order to test the performance of electro-optic devices such as high-speed electro-optic modulator, we took several years building a high-speed test platform which can characterize high-speed transmission spectrum, frequency response and eye diagrams. The schematic diagrams of these platforms are shown in Figs. 2.26, 2.30 and 2.33, respectively.

First, we tested the transmission spectrum of the modulator sample. A tunable laser (Agilent 8164B, 81600-201) is used as the light source. We used a polarization controller to change the polarization of the input light, and used a lensed fiber to couple the light into the waveguide. The light passed through a lensed fiber and is finally detected in a optical power-meter (Agilent 81624B). The whole link is shown in Fig. 2.26b.

Figures 2.27 and 2.28 show the normalized spectrum when the reverse bias voltage changes from 0 V to 8 V for the phase shifter length of 2 mm and 1 mm in the modulators, respectively. In order to get the actual insertion loss of the modulator, we first use the cut-back method to calculate the propagation loss of the passive waveguides by 5 dB/cm. After a series of calculations, the modulator samples with a length of 2 mm and 1 mm can be calculated with a insertion loss of 6.2 dB and 3.9 dB, respectively. It can also be calculated that the loss of the phase shifter is 2.8 dB/mm, and the excess loss of the MMI coupler is 0.35 dB. The loss of the phase shifter can be improved by reducing the doping concentration. However, this will destroy the

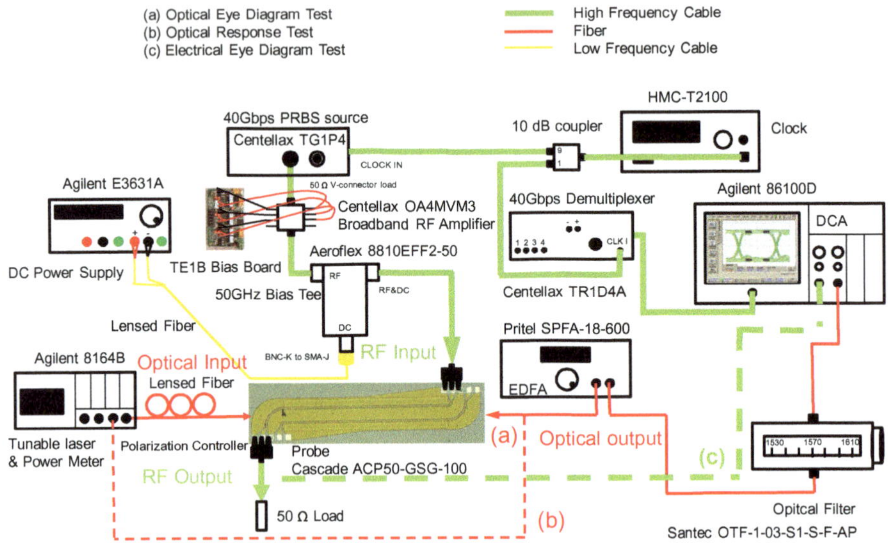

Fig. 2.26 44 Gb/s high-speed test set-up in SIMIT, CAS

Fig. 2.27 Normalized Transmission spectrum of the modulator with 2 mm long phase shifter under reverse bias from 0 to 8 V. [2013] IEEE. Reprinted, with permission, from Ref. [75]

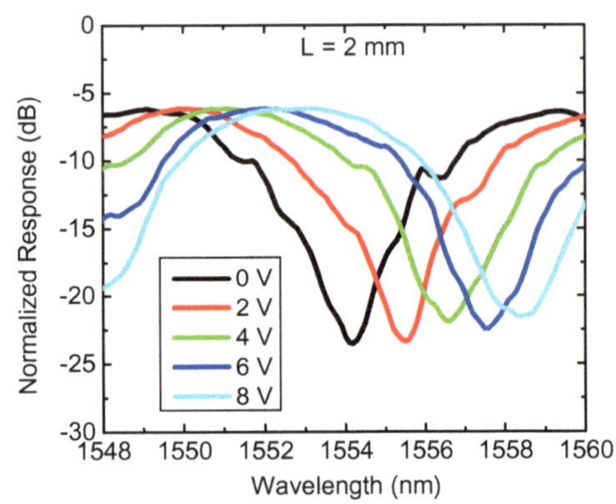

modulation efficiency of $V_\pi L_\pi$. In our previous work, we realized a MMI with 0.06 dB excess loss [79] and a low-loss optical waveguide with of 2.4 dB/cm propagation loss [80] by optimizing the process. Therefore, it is assumed that the passive loss of the modulator can be optimized to the above extent. For the modulator sample with a phase shifter length of 1 mm, the insertion loss can be reduced to 2.8 dB without affecting its modulation efficiency.

According to the spectra shift, we can further estimate the modulation efficiency of the modulator $V_\pi L_\pi$. Figure 2.29 shows the modulation efficiency $V_\pi L_\pi$ at different bias voltages. For the modulator samples with a phase shifter length of 2 mm and 1 mm, the modulation efficiency are 1.62–2.05 V·cm and 1.47–1.97 V·cm, respectively. As the PN junction doping concentration in the actual process may be higher than

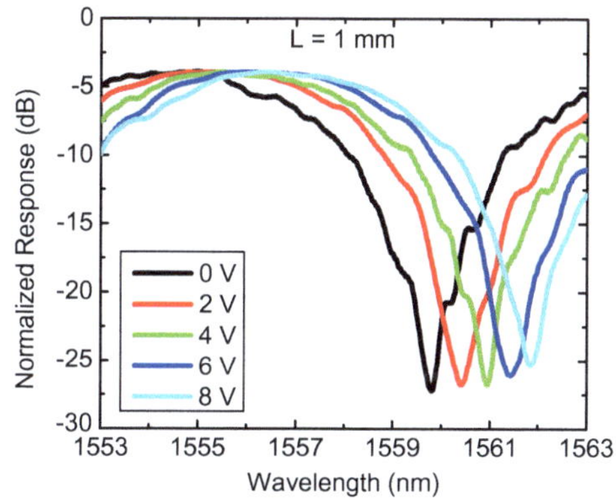

Fig. 2.28 Normalized Transmission spectrum of the modulator with 1 mm long phase shifter under reverse bias from 0 to 8 V. [2013] IEEE. Reprinted, with permission, from Ref. [75]

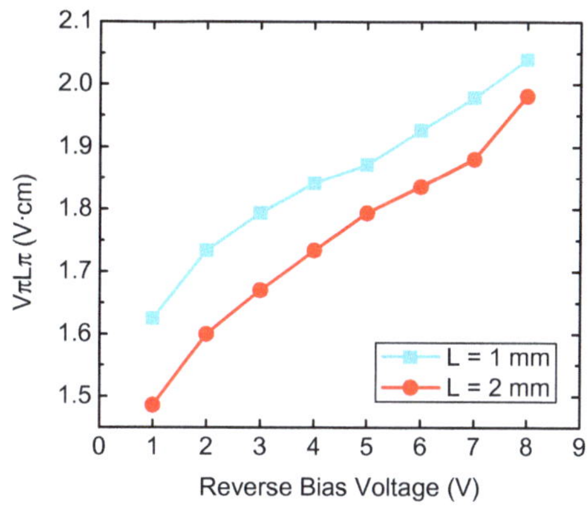

Fig. 2.29 Modulation efficiency $V_{\pi}L_{\pi}$ as a function of reverse bias voltage. [2013] IEEE. Reprinted, with permission, from Ref. [75]

that in the simulation, the measured modulation efficiency will be slightly better than the simulation value $V_{\pi}L_{\pi}$. For a longer device and a smaller driving voltage, a better modulation efficiency $V_{\pi}L_{\pi}$ can be obtained, which can be explained as that the depletion width of PN junction is proportional to the square root of the driving voltage [81, 82].

The S parameters of the modulator electrodes can be measured by using a vector network analyzer (VNA, Agilent N5244A) operating from 100 MHz to 43.5 GHz. The vector network analyzer is connected with the sample by high frequency cables (Agilent 85133F and Cascade 103-202-B) and 50 GHz GSG high frequency probes (Cascade ACP50-GSG-100). The DC bias is controlled by the voltage source (Agilent E3631A) connected to the corresponding port of the network analyzer, so there is no need for an additional bias-tee. The whole test link is shown in Fig. 2.30. Before the test, we used the impedance standard substrate to carry out a calibration (two-port short-open-load-through (SOLT)) calibration to remove the influence of cable and probe on the S parameter measurement. So the measured S parameter is only caused by the modulator electrodes.

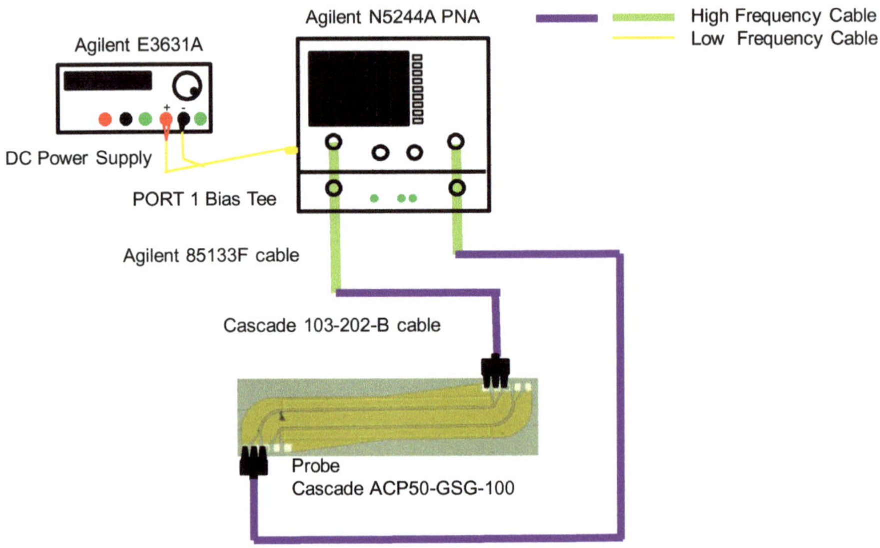

Fig. 2.30 Test set-up for characterizing electrical S parameters in SIMIT, CAS

Fig. 2.31 Normalized electrical S21 parameter of the modulator with 2 mm long phase shifter under reverse bias from 0 to 5 V. [2013] IEEE. Reprinted, with permission, from Ref. [75]

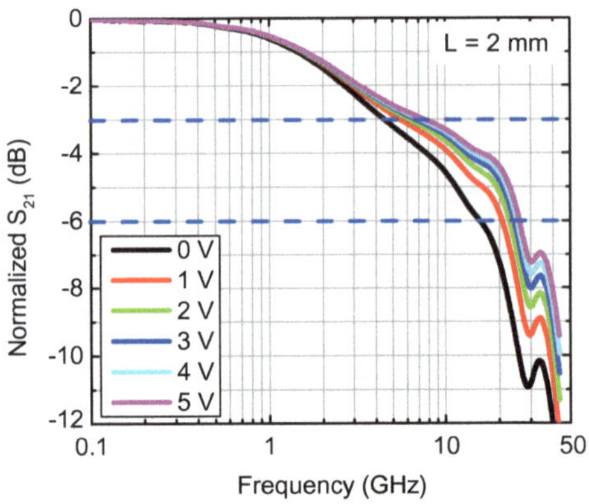

Figures 2.31 and 2.32 show the normalized S21 parameter of the modulator sample with the phase shifter length of 2 mm and 1 mm at the bias voltage from 0 V to –5 V, respectively. With the bias voltage increasing, the bandwidth will increase. This is because as the depletion width increases, the junction capacitance decreases. For a modulator sample with a 2 mm-long phase shifter, when the bias voltage are 0 V and –5 V, the 6 dB bandwidth are 16 GHz and 26 GHz, respectively. For the modulator sample with a phase shifter length of 1 mm, when the bias voltage is –2 V, the 6 dB bandwidth is more than 40 GHz, which is good enough to support a high-speed transmission up to 40 Gb/s.

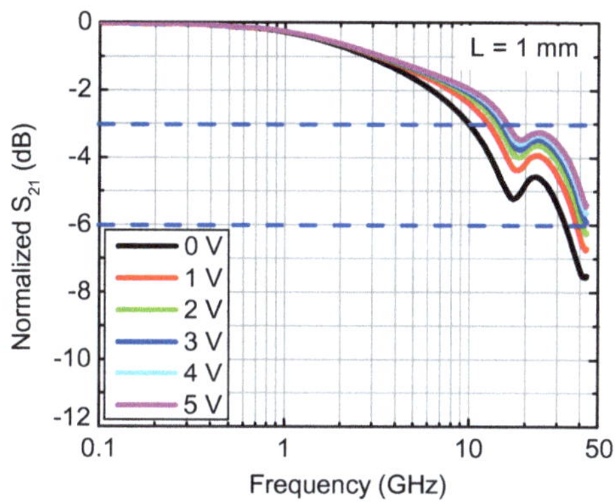

Fig. 2.32 Normalized electrical S21 parameter of the modulator with 1 mm long phase shifter under reverse bias from 0 to 5 V. [2013] IEEE. Reprinted, with permission, from Ref. [75]

The electro-optic spectral response of modulator samples can be tested by combining VNA (Agilent N5244A) and LCA (Agilent N4373C). After the light from the 1550 nm light source in the LCA enters into the device, it is then returned to the LCA through an erbium doped fiber amplifier (EDFA, PriTel SPFA-18-600) and an tunable optical filter (Santec OTF-1-03-S1-S-F-AP). One end of the modulator electrode is connected to the vector network analyzer through a 50 GHz GSG high-frequency probe (Cascade ACP50-GSG-100), and the other end is connected by a high-frequency probe to a 50 GHz DC block and a 50 ohm load. The whole test link is shown in Fig. 2.33.

Figure 2.34 shows the electro-optic frequency response of the modulator with different lengths when the bias is –2 V. There are some ripples in the frequency response spectrum, which may be due to the failure of probe calibration. We normalized the frequency response. For the modulator samples with arm length of 2 mm and 1 mm, the measured 3 dB bandwidth are 30 GHz and 19 GHz, respectively. We also showed the simulation results as a comparison, and we can find that the simulated 3 dB bandwidth are 33 GHz and 22 GHz respectively, which agreed well with the measurement results.

The large-signal performance of modulator is mainly evaluated by testing the high-speed optical eye diagram. In the driving circuit part, a 40 Gb/s pseudo-random binary signal with 2^{31}-1 length needs to be generated by using a clock source (Hittite HMC-T2100), a 40 Gb/s signal generator (Centellax TG1P4), a 40 Gb/s demultiplexer (Centellax TR1D4A) and a series of high-frequency components. The high-speed signal then enters into the electrode of the modulator through a broadband radio-frequency amplifier (Centellax OA4MVM3) and a 50 GHz bias-tee (Aeroflex 8810EFF2-50). The output end of the electrode is connected to a 50 ohm terminator. In the optical part, a tunable laser (Agilent 8164B, 81600-201) is used as a light source. The optical signal is modulated by a optical modulator and then go through an erbium-doped fiber amplifier (EDFA, PriTel, SPFA-18-600) and a tunable optical filter (Santec OTF-1-03-S1-S-F-AP), which is finally received by a high-speed digital communication analyzer (DCA, Agilent 86100D and 86116C-040). The overall link is shown in Fig. 2.26a.

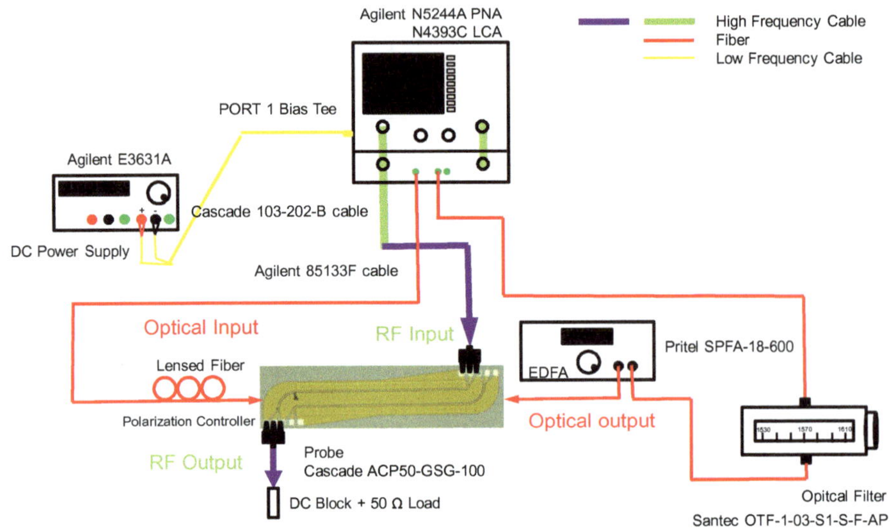

Fig. 2.33 43.5 GHz small signal test set-up in SIMIT, CAS

Fig. 2.34 Small signal test result of modulator with different lengths under 2 V reverse bias voltage. [2013] IEEE. Reprinted, with permission, from Ref. [75]

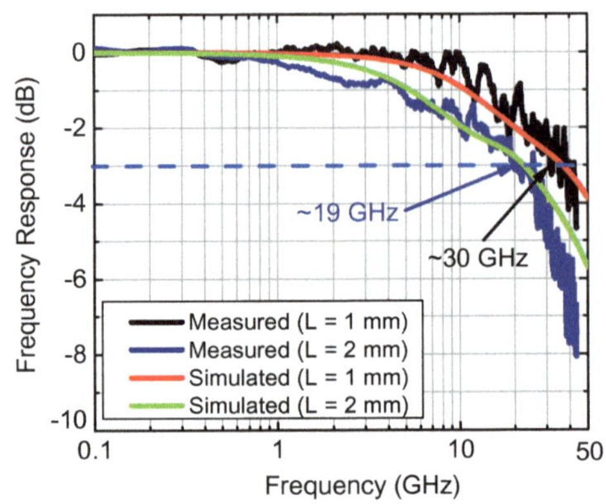

Figure 2.35a–b show the high-speed eye diagrams of the modulator with 2 mm and 1 mm phase shifter, respectively. For a 2 mm modulator sample, when the peak-to-peak voltage $V_{pp} = 3$ V and the bias voltage $V_{bias} = -5$ V, a 35 Gb/s optical eye can be obtained, and its extinction ratio is 5.84 dB. In the case of the same bias voltage V_{bias} and higher $V_{pp} = 5$ V, a 44 Gb/s optical eye with 2.44 dB extinction ratio is obtained in the 1 mm long modulator.

In order to evaluate the practical transmission performance of the modulator, we tested the bit error rate of these samples with 30 Gb/s data rate, as shown in Fig. 2.36. It can be found that the 1 mm-long modulator sample exhibits better transmission performances. And the error-free transmission can be obtained in these two samples (defined as $BER < 5 \times 10^{-10}$), but the longer sample requires higher input power.

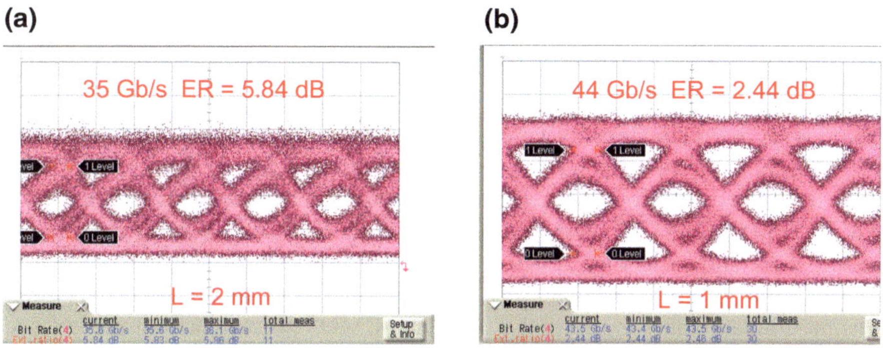

Fig. 2.35 High-speed optical eye diagram of modulator with different lengths. **a** 2 mm, 35 Gb/s. **b** 1 mm, 44 Gb/s. [2013] IEEE. Reprinted, with permission, from Ref. [75]

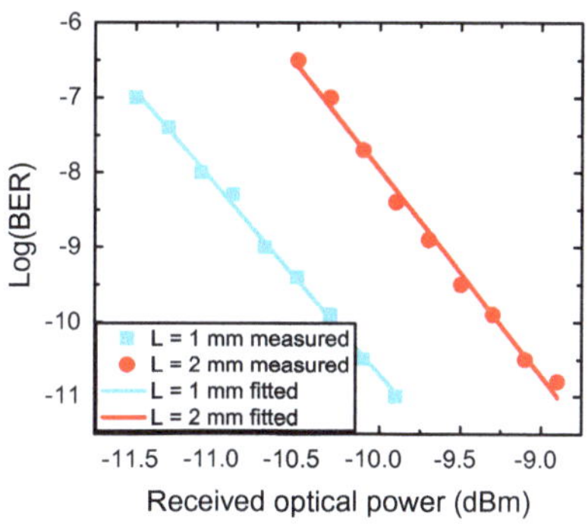

Fig. 2.36 Bit-error rate of the modulators operating at 30 Gb/s. [2013] IEEE. Reprinted, with permission, from Ref. [75]

In addition, we also calculated the power consumption of these two modulator samples. A common method of estimating power is to use the formula introduced in the Ref. [83].

$$P = \frac{{V_{\text{pp}}}^2}{4 \times \text{BR} \times Z} \tag{2.81}$$

Here, BR is the transmission rate and Z is the system impedance (50 Ohm normally). By using Eq. 2.81, we can calculate the power consumption is 2.84 pJ/bit in a 1 mm-long modulator when driving voltage $V_{\text{pp}} = 5$ V and transmission speed is 44 Gb/s. In order to reduce the power consumption, we can use a longer phase shifter [84] and optimize the PN junction carefully [85].

2.6 30 Gb/s Carrier Depletion Silicon Electro-Optic Modulator Based on MRR

Some MRR optical modulators were fabricated in the same 130 nm CMOS processing platform. The microscope picture can be found in Fig. 2.37.

In the first process flow, we fabricated the modulator samples with coupling gap of 120 nm, 150 nm, and 170 nm, respectively. Their transmission spectrum measurement results were shown in Fig. 2.38a–d. In this process, we did not find the normal transmission spectrum of MRR. And when the gap is 120 nm, the loss is very high. The preliminary analysis shows that a 130 nm CMOS process can not achieve such a small coupling in the modulator sample, so the bus waveguide and the microring have been linked together, causing the light unable to guide.

Figure 2.39a–d show the measured transmission spectrum of the MRR optical modulator samples in the second process flow. Because of the limited time, the layout has not been modified, but some adjustments have been made for the process. Compared to the previous chip, it is possible to improve the fabrication process and some light can be observed for the case of 120 nm gap, but there is still no resonance spectrum. When the gap is 170 nm, the resonance can be obviously seen. The FSR is about 16.27 nm with a low extinction ratio of 6 dB at the wavelength of 1553.28 nm, and the jitter is a little large.

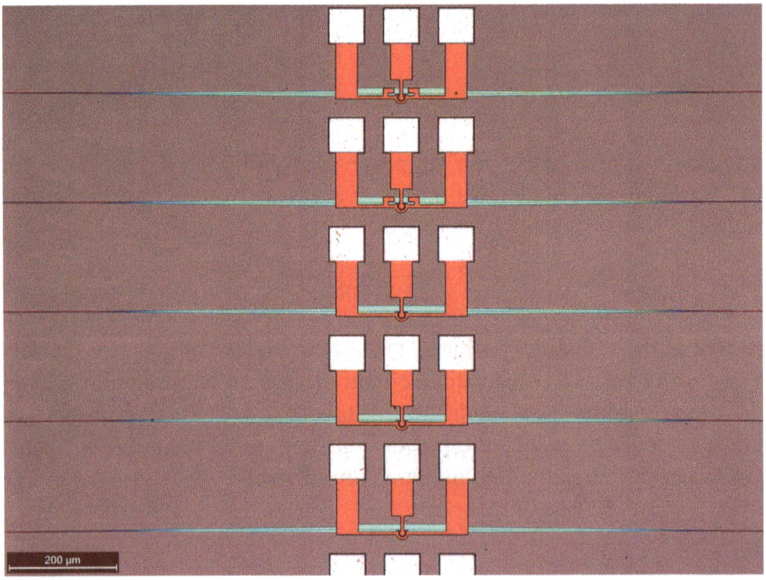

Fig. 2.37 Microscoped picture of MRR electro-optical modulator sample

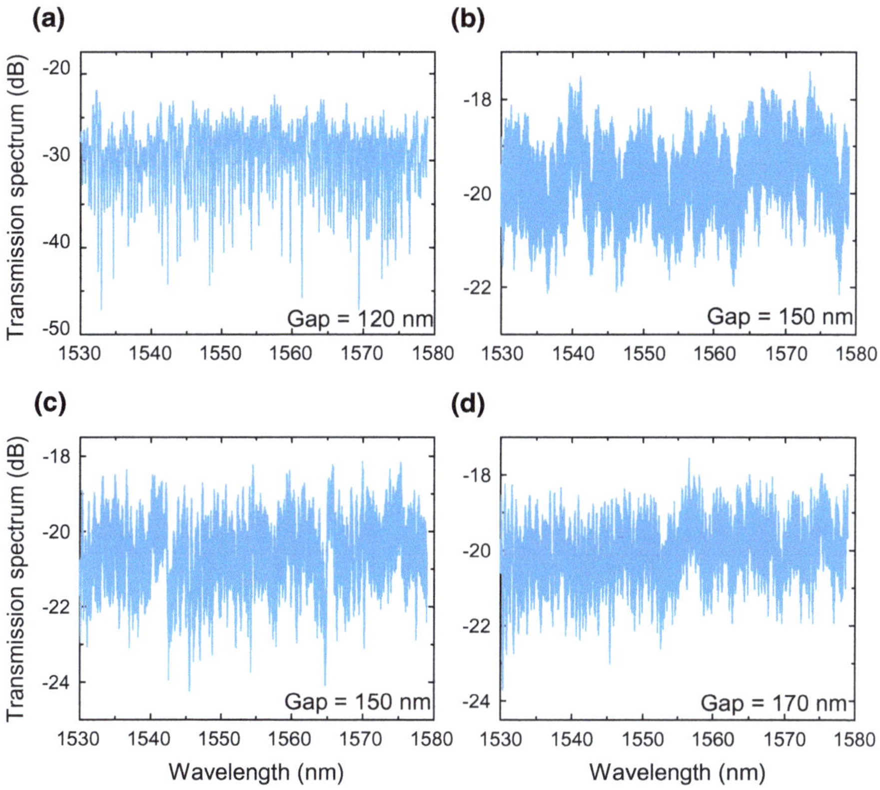

Fig. 2.38 Measured transmission spectrum of MRR modulator in the first process

Figure 2.40a–d show the measured transmission spectrum of the MRR electro-optic modulator samples in the third processing. According to the test results of the first two processing, we abandoned the design of the modulator with 120 nm gap, and added one with 200 nm gap. From the test results, it can be found that the jitter is significantly reduced with smaller than 1 dB. And it can be seen that the extinction ratio of the sample with 200 nm gap can be 12 dB. According to Eq. 2.42, we can get the Q factor of about 4900.

We finally selected the modulator sample with 200 nm gap in the third processing. The microscope picture is as shown in Fig. 2.41. Its size is very small and the active region is only $30 \times 20\ \mu m^2$.

The performances of the modulator sample were characterized in our testing platform. Figure 2.42 shows the transmission spectrum of the MRR electro-optic modulator samples under different bias voltages. It can be found that by adding an reverse bias voltage, the resonance wavelength of MRR is red-shifted and the red shift is more obvious when the reverse bias voltage further decreases. The main reason is

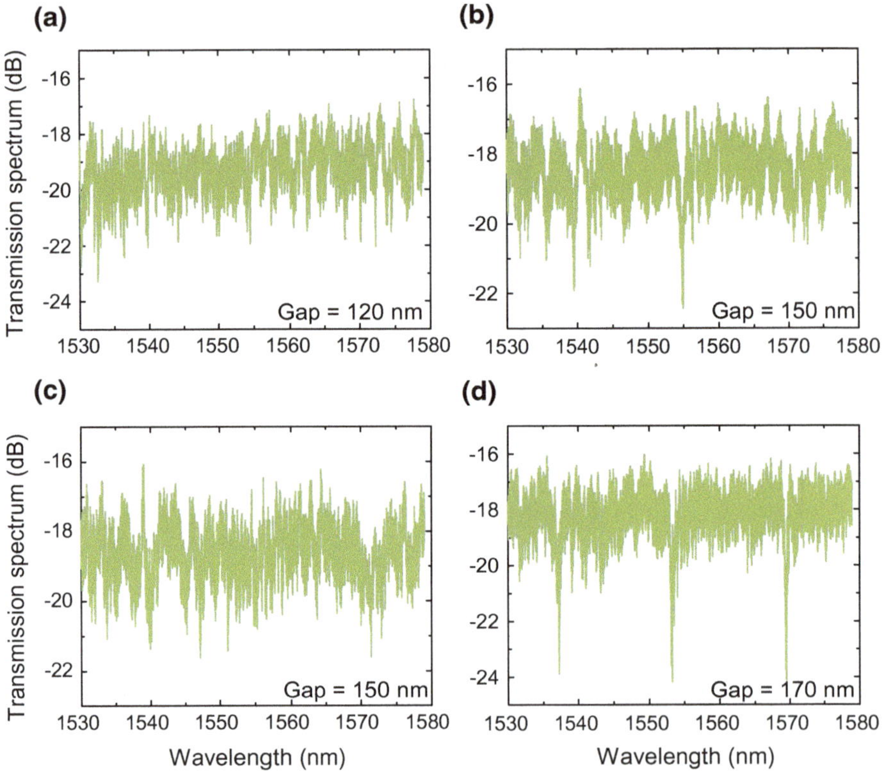

Fig. 2.39 Measured transmission spectrum of MRR modulator in the second process

that with the decreasing of the reverse bias voltage, the carrier concentration in the active region decreases, resulting in higher refractive index.

We further extracted the modulation efficiency $V_\pi L_\pi$ of the modulator samples at different voltages, as shown in Fig. 2.43. When the reverse bias is between –1 and –6 V, the modulation efficiency $V_\pi L_\pi$ arises from 0.87 to 1.77 V·cm. The modulation efficiency increases with the decreasing of the voltage because the width of the depletion layer is proportional to the square root of the driving voltage.

In addition, we have also tested the optical eye diagram of the modulator at different rates, as shown in Fig. 2.44. We used a new waveform analysis software in the DCA, so the color of the eye is not as the same as that of the MZI modulator. When the peak-to-peak voltage is $V_{pp} = 4.5$ V, the bias voltage is $V_{bias} = -4.1$ V and the wavelength is 1554.88 nm, a clear optical eye diagram with the maximum rate of 30 Gb/s can be obtained.

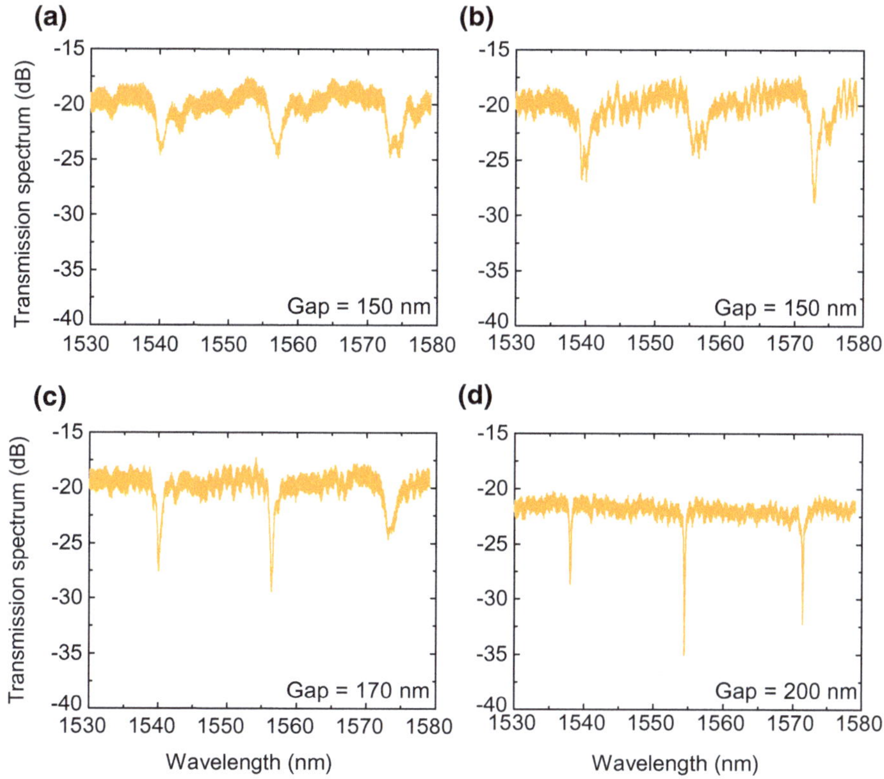

Fig. 2.40 Measured transmission spectrum of MRR modulator in the third process

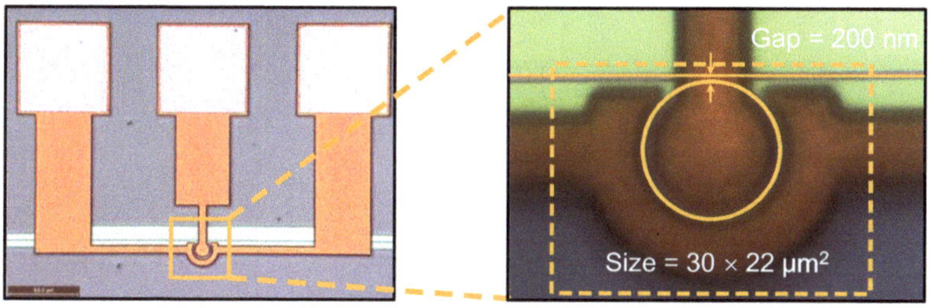

Fig. 2.41 Microscope picture of the MRR modulator with 200 nm gap in the third process

In order to evaluate the power consumption of the MRR electro-optic modulator, first we used a vector network analyzer to test the S11 parameter of the modulator, and fitted the test results with the circuit parameters model of the MRR modulator, as shown in Fig. 2.45. The PN-junction capacitance C_j is 21 fF, and the pad capacitance

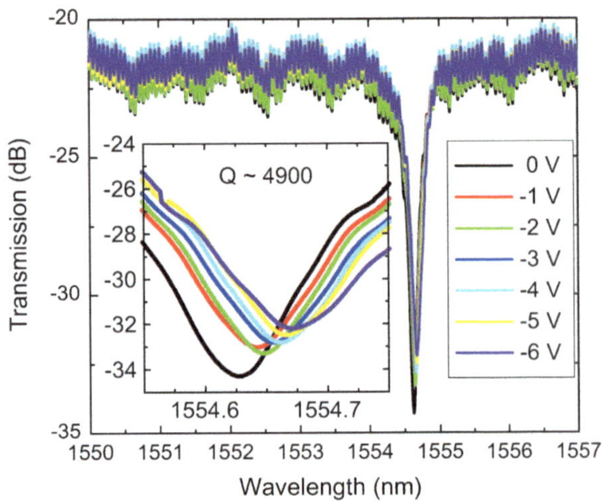

Fig. 2.42 Measured transmission spectrum of MRR modulator under different reverse bias voltage

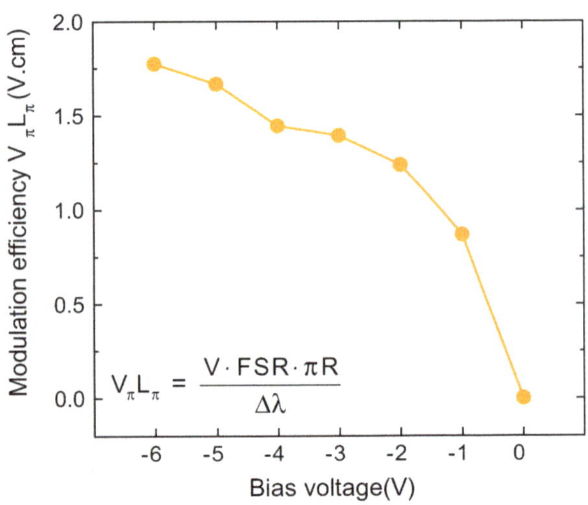

Fig. 2.43 Modulation efficiency $V_\pi L_\pi$ as a function of bias voltage in a MRR optical modulator

C_p is 12 fF. Finally, the power consumption $P = (21+12) \times 4.5^2/4 = 167$ fJ/bit according to the relationship $P = C{\cdot}V^2/4$. To further reduce the power consumption, we can reduce the driving voltage.

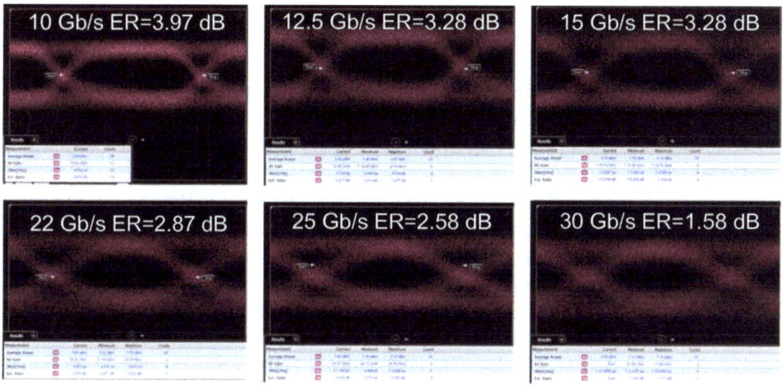

Fig. 2.44 Optical eye diagram with different speed in a MRR modulator

Fig. 2.45 S11 parameter test and fitting result of the MRR modulator. Insets: fitting equivalent circuit model

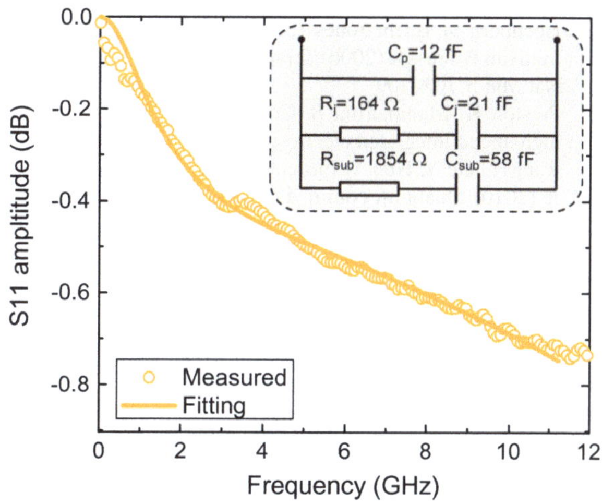

2.7 Conclusion

In this chapter, we first introduced the operation principle of the silicon electro-optic modulator, the plasma dispersion effect, and reviewed and summarized the research progress in the past few decades. Then, we put forward the design theory of silicon electro-optic modulator systematically, and introduced the method of optimizing the waveguide size by combining the semiconductor process simulation software and in-house programming. Then, we studied the equivalent circuit model of the modulator with lateral PN junction structure and the extraction of each circuit element in details. The travelling-wave electrode is optimized by using the theory of microwave transmission line. Finally, we used the standard CMOS technology to demonstrate a 44 Gb/s MZI silicon modulator and a 30 Gb/s low power-consumption MRR silicon modulator.

References

1. Gan F (2007) High-speed silicon electro-optic modulator for electronic photonic integrated circuits. PhD diss, Massachusetts Institute of Technology
2. Soref RA, Bennett BR (1987) Electrooptical effects in silicon. IEEE J Quantum Electron 23:123–129
3. Lim AEJ, Song J, Fang Q, Li C, Tu X, Duan N, Liow TY (2014) Review of silicon photonics foundry efforts. IEEE J Sel Top Quantum Electron 20:405–416
4. Rickman A (2014) The commercialization of silicon photonics. Nat Photon 8:579–582
5. Reed GT, Mashanovich GZ, Gardes FY, Nedeljkovic M, Hu Y, Thomson DJ, Li K, Wilson PR, Chen SW, Hsu SS (2014) Recent breakthroughs in carrier depletion based silicon optical modulators. Nanophotonics 3:229–245
6. Reed GT, Mashanovich G, Gardes FY, Thomson DJ (2010) Silicon optical modulators. Nat Photonics 4:518–526
7. Jacobsen RS, Andersen KN, Borel PI, Fage-Pedersen J, Frandsen LH, Hansen O, Kristensen M, Lavrinenko AV, Moulin G, Ou H, Peucheret C (2006) Strained silicon as a new electro-optic material. Nature 441:199–202
8. Hochberg M, Baehr-Jones T, Wang G, Shearn M, Harvard K, Luo J, Chen B, Shi Z, Lawson R, Sullivan P, Jen AK (2006) Terahertz all-optical modulation in a silicon–polymer hybrid system. Nat Mat 5:703–709
9. Preston K, Manipatruni S, Gondarenko A, Poitras CB, Lipson M (2009) Deposited silicon high-speed integrated electro-optic modulator. Opt Express 17:5118–5124
10. Rong Y, Ge Y, Huo Y, Fiorentino M, Tan MR, Kamins TI, Ochalski TJ, Huyet G, Harris JS Jr (2010) Quantum-confined Stark effect in Ge/SiGe quantum wells on Si. IEEE J Sel Top Quantum Electron 16:85–92
11. Chen HW, Kuo YH, Bowers JE (2008) High speed hybrid silicon evanescent Mach-Zehnder modulator and switch. Opt Express 16:20571–20576
12. Chen HW, Kuo YH, Bowers JE (2008) A Hybrid silicon-AlGaInAs phase modulator. IEEE Photonics Technol Lett 20:1920–1922
13. Chen HW, Peters JD, Bowers JE (2011) Forty Gb/s hybrid silicon Mach-Zehnder modulator with low chirp. Opt Express 19:1455–1460
14. Tang Y, Chen HW, Jain S, Peters JD, Westergren U, Bowers JE (2011) 50 Gb/s hybrid silicon traveling-wave electroabsorption modulator. Opt Express 19:5811–5816
15. Tang Y, Peters JD, Bowers JE (2012) Energy-efficient hybrid silicon electroabsorption modulator for 40 Gb/s 1 V uncooled operation. IEEE Photonics Technol Lett 24:1689–1692
16. Tang Y, Peters JD, Bowers JE (2012) Over 67 GHz bandwidth hybrid silicon electroabsorption modulator with asymmetric segmented electrode for 1.3 μ m transmission. Opt Express 20:11529–11535
17. Liu J, Beals M, Pomerene A, Bernardis S, Sun R, Cheng J, Kimerling LC, Michel J (2008) Waveguide-integrated, ultralow-energy GeSi electro-absorption modulators. Nat Photonics 2:433–437
18. Lim PH, Cai J, Ishikawa Y, Wada K (2012) Laterally coupled silicon-germanium modulator for passive waveguide systems. Opt Express 37:1496–1498
19. Feng D, Liao S, Liang H, Fong J, Bijlani B, Shafiiha R, Luff B, Luo Y, Cunningham J, Krishnamoorthy AV, Asghari M (2012) High speed GeSi electro-absorption modulator at 1550 nm wavelength on SOI waveguide. Opt Express 20:22224–22232
20. Edwards EH, Lever L, Fei ET, Kamins TI, Ikonic Z, Harris JS, Kelsall RW, Miller DA (2013) Low-voltage broad-band electroabsorption from thin Ge/SiGe quantum wells epitaxially grown on silicon. Opt Express 21:867–876
21. Wu P, Clarke RE, Novak J, Deng S, McDonald JF, Huang ZR (2013) Ultrashort SiGe heterojunction bipolar transistor-based high-speed optical modulator. IEEE J Sel Top Quantum Electron 19:7900109–7900109

22. Leuthold J, Koos C, Freude W, Alloatti L, Palmer R, Korn D, Pfeifle J, Lauermann M, Dinu R, Wehrli S, Jazbinsek, M (2013) Silicon-organic hybrid electro-optical devices. IEEE J Sel Top Quantum Electron 19:114–126
23. Alloatti L, Palmer R, Diebold S, Pahl KP, Chen B, Dinu R, Fournier M, Fedeli JM, Zwick T, Freude W, Koos C (2014) 100 GHz silicon–organic hybrid modulator. Light Sci Appl 3:1–4
24. Koeber S, Palmer R, Lauermann M, Heni W, Elder DL, Korn D, Woessner M, Alloatti L, Koenig S, Schindler PC, Yu, H (2014) Femtojoule electro-optic modulation using a silicon-organic hybrid device. Light Sci Appl 4:1–8
25. Gould M, Baehr-Jones T, Ding R, Huang S, Luo J, Jen AKY, Fedeli JM, Fournier M, Hochberg M (2011) Silicon-polymer hybrid slot waveguide ring-resonator modulator. Opt Express 19:3952–3961
26. Liu M, Yin X, Ulin-Avila E, Geng B, Zentgraf T, Ju L, Wang F, Zhang X (2011) A graphene-based broadband optical modulator. Nature 474:64–67
27. Melikyan A, Alloatti L, Muslija A, Hillerkuss D, Schindler PC, Li J, Palmer R, Korn D, Muehlbrandt S, Van Thourhout D, Chen B (2014) High-speed plasmonic phase modulators. Nat Photonics 8:229–233
28. Soref RA, Bennett BR (1987) Kramers-Kronig analysis of electro-optical switching in silicon. SPIE Integr Opt Circuit Eng 704:32–37
29. Manipatruni S, Xu Q, Schmidt B, Shakya J, Lipson M (2007) High speed carrier injection 18 Gb/s silicon micro-ring electro-optic modulator. In: IEEE proceedings for the lasers and electro-optics society, vol 537–538
30. Manipatruni S, Xu Q, Schmidt B, Shakya J, Lipson M (2004) A high-speed silicon optical modulator based on a metal-oxide-semiconductor capacitor. Nature 427:615–618
31. Gardes FY, Reed GT, Emerson NG, Png CE (2005) A sub-micron depletion-type photonic modulator in silicon on insulator. Opt Express 13:8845–8854
32. Liu A, Liao L, Rubin D, Nguyen H, Ciftcioglu B, Chetrit Y, Izhaky N, Paniccia M (2007) High-speed optical modulation based on carrier depletion in a silicon waveguide. Opt Express 15:660–668
33. Liao L, Liu A, Rubin D, Basak JA, Chetrit YA, Nguyen HA, Cohen RA, Izhaky NA, Paniccia MA (2007) 40 Gbit/s silicon optical modulator for high-speed applications. Electron Lett 43:1196–1197
34. Dong P, Chen L, Chen YK (2012) High-speed low-voltage single-drive push-pull silicon Mach-Zehnder modulators. Opt Express 20:6163–6169
35. Thomson DJ, Gardes FY, Fedeli JM, Zlatanovic S, Hu Y, Kuo BP, Myslivets E, Alic N, Radic S, Mashanovich GZ, Reed GT (2012) 50 Gb/s silicon optical modulator. IEEE Photonics Technol Lett 24:234–236
36. You JB, Park M, Park JW, Kim G (2008) 12.5 Gbps optical modulation of silicon racetrack resonator based on carrier-depletion in asymmetric pn diode. Opt Express 16:18340–18344
37. Xiao X, Xu H, Li X, Li Z, Chu T, Yu Y, Yu J (2013) High-speed, low-loss silicon Mach-Zehnder modulators with doping optimization. Opt Express 21:4116–4125
38. Yi H, Long Q, Tan W, Li L, Wang X, Zhou Z (2012) Demonstration of low power penalty of silicon Mach-Zehnder modulator in long-haul transmission. Opt Express 20:27562–27568
39. Gardes FY, Brimont A, Sanchis P, Rasigade G, Marris-Morini D, O'Faolain L, Dong F, Fedeli JM, Dumon P, Vivien L, Krauss TF (2009) High-speed modulation of a compact silicon ring resonator based on a reverse-biased pn diode. Opt Express 17:21986–21991
40. Rasigade G, Marris-Morini D, Vivien L, Cassan E (2010) Performance evolutions of carrier depletion silicon optical modulators: from PN to PIPIN diodes. IEEE J Sel Top Quantum Electron 16:179–184
41. Rasigade G, Marris-Morini D, Ziebell M, Cassan E, Vivien L (2011) Analytical model for depletion-based silicon modulator simulation. Opt Express 19:3919–3924
42. Marris-Morini D, Vivien L, Fédéli JM, Cassan E, Lyan P, Laval S (2008) Low loss and high speed silicon optical modulator based on a lateral carrier depletion structure. Opt Express 16:334–339

43. Rasigade G, Ziebell M, Marris-Morini D, Fédéli JM, Milesi F, Grosse P, Bouville D, Cassan E, Vivien L (2011) High extinction ratio 10 Gbit/s silicon optical modulator. Opt Express 19:5827–5832
44. Gutierrez AM, Brimont A, Rasigade G, Ziebell M, Marris-Morini D, Fedeli JM, Vivien L, Marti J, Sanchis P (2012) Ring-assisted Mach-Zehnder interferometer silicon modulator for enhanced performance. J Light Technol 30:9–14
45. Rasigade G, Ziebell M, Marris-Morini D, Brimont A, Campo AG, Sanchis P, Fédéli JM, Duan GH, Cassan E, Vivien L: 10 Gb/s error-free silicon optical modulator for both TE and TM polarized light. J Light Technol 23:1799–1801
46. Marris-Morini D, Baudot C, Fédéli JM, Rasigade G, Vulliet N, Souhaité A, Ziebell M, Rivallin P, Olivier S, Crozat P, Le Roux X (2013) Low loss 40 Gbit/s silicon modulator based on interleaved junctions and fabricated on 300 mm SOI wafers. Opt Express 21:22471–22475
47. Tu X, Liow TY, Song J, Yu M, Lo GQ (2011) Fabrication of low loss and high speed silicon optical modulator using doping compensation method. Opt Express 19:18029–18035
48. Tu X, Liow TY, Song J, Luo X, Fang Q, Yu M, Lo GQ (2013) 50 Gb/s silicon optical modulator with traveling-wave electrodes. Opt Express 21:12776–12782
49. Yang Y, Fang Q, Yu M, Tu X, Rusli R, Lo GQ (2014) High-efficiency Si optical modulator using Cu travelling-wave electrode. Opt Express 22:29978–29985
50. Thomson DJ, Gardes FY, Reed GT, Milesi F, Fedeli JM (2010) High speed silicon optical modulator with self aligned fabrication process. Opt Express 18:19064–19069
51. Thomson DJ, Gardes FY, Reed GT, Milesi F, Fedeli JM (2011) 40 Gb/s silicon photonics modulator for TE and TM polarisations. Opt Express 19:11804–11814
52. Thomson DJ, Gardes FY, Liu S, Porte H, Zimmermann L, Fedeli JM, Hu Y, Nedeljkovic M, Yang X, Petropoulos P, Mashanovich GZ (2013) High performance Mach-Zehnder-based silicon optical modulators. IEEE J Sel Top Quantum Electron 19:85–94
53. Thomson DJ, Gardes FY, Hu Y, Mashanovich G, Fournier M, Grosse P, Fedeli JM, Reed GT (2011) High contrast 40 Gbit/s optical modulation in silicon. Opt Express 19:11507–11516
54. Li ZY, Xu DX, McKinnon WR, Janz S, Schmid JH, Cheben P, Yu JZ (2009) Silicon waveguide modulator based on carrier depletion in periodically interleaved PN junctions. Opt Express 17:15947–15958
55. Hu Y, Xiao X, Xu H, Li X, Xiong K, Li Z, Chu T, Yu Y, Yu J (2012) High-speed silicon modulator based on cascaded microring resonators. Opt Express 20:15079–15085
56. Xiao X, Xu H, Li X, Hu Y, Xiong K, Li Z, Chu T, Yu Y, Yu J (2012) 25 Gbit/s silicon microring modulator based on misalignment-tolerant interleaved PN junctions. Opt Express 20:2507–2515
57. Xu H, Xiao X, Li X, Hu Y, Li Z, Chu T, Yu Y, Yu J (2012) High speed silicon Mach-Zehnder modulator based on interleaved PN junctions. Opt Express 20:15093–15099
58. Hao X, Xian-Yao L, Xi X, Zhi-Yong L, Yu-De Y, Jin-Zhong Y (2013) High-speed and broad optical bandwidth silicon modulator. Chin Phys B 22:114212–114216
59. Xu H, Li X, Xiao X, Li Z, Yu Y, Yu J (2014) Demonstration and characterization of high-speed silicon depletion-mode Mach-Zehnder modulators. IEEE J Sel Top Quantum Electron 20:23–32
60. Yu H, Pantouvaki M, Van Campenhout J, Korn D, Komorowska K, Dumon P, Li Y, Verheyen P, Absil P, Alloatti L, Hillerkuss D (2012) Performance tradeoff between lateral and interdigitated doping patterns for high speed carrier-depletion based silicon modulators. Opt Express 20:12926–12938
61. Pantouvaki M, Yu H, Rakowski M, Christie P, Verheyen P, Lepage G, Van Hoovels N, Absil P, Van Campenhout J (2013) Comparison of silicon ring modulators with interdigitated and lateral PN junctions. IEEE J Sel Top Quantum Electron 19:7900308–7900308
62. Ziebell M, Marris-Morini D, Rasigade G, Crozat P, Fédéli JM, Grosse P, Cassan E, Vivien L (2011) Ten Gbit/s ring resonator silicon modulator based on interdigitated PN junctions. Opt Express 19:14690–14695
63. Xiao X, Li X, Xu H, Hu Y, Xiong K, Li Z, Chu T, Yu J, Yu Y (2012) 44-Gb/s silicon microring modulators based on zigzag PN junctions. Opt Express 24:1712–1714

64. Spector SJ, Sorace CM, Geis MW, Grein ME, Yoon JU, Lyszczarz TM, Kartner FX (2010) Operation and optimization of silicon-diode-based optical modulators. IEEE J Sel Top Quantum Electron 16:165–172
65. Liow TY, Song J, Tu X, Lim AE, Fang Q, Duan N, Yu M, Lo GQ (2013) Silicon optical interconnect device technologies for 40 Gb/s and beyond. IEEE J Sel Top Quantum Electron 19:8200312–8200312
66. Rosenberg JC, Green WM, Assefa S, Gill DM, Barwicz T, Yang M, Shank SM, Vlasov YA (2012) A 25 Gbps silicon microring modulator based on an interleaved junction. Opt Express 20:26411–26423
67. Gill DM, Patel SS, Rasras M, Tu KY, White AE, Chen YK, Pomerene A, Carothers D, Kamocsai RL, Hill CM, Beattie J (2010) CMOS-compatible Si-ring-assisted Mach-Zehnder interferometer with internal bandwidth equalization. J IEEE J Sel Top Quantum Electron 16:45–52
68. Li X, Xiao X, Xu H, Li Z, Chu T, Yu J, Yu Y (2013) Highly efficient silicon Michelson interferometer modulators. IEEE J Sel Top Quantum Electron 25:407–409
69. Patel D, Veerasubramanian V, Ghosh S, Samani A, Zhong Q, Plant DV (2014) Highly efficient silicon Michelson interferometer modulators. Opt Express 22:26788–26802
70. Yu H, Pantouvaki M, Dwivedi S, Verheyen P, Lepage G, Baets R, Bogaerts W, Absil P, Van Campenhout J (2013) Compact thermally tunable silicon racetrack modulators based on an asymmetric waveguide. IEEE Photonics Technol Lett 25:159–162
71. Timurdogan E, Sorace-Agaskar CM, Sun J, Hosseini ES, Biberman A, Watts MR (2014) An ultralow power athermal silicon modulator. Nat Commun 5:1–11
72. Rabus DG (2014) Integrated ring resonators. Springer, Berlin
73. Li G, Krishnamoorthy AV, Shubin I, Yao J, Luo Y, Thacker H, Zheng X, Raj K, Cunningham JE (2013) Ring resonator modulators in silicon for interchip photonic links. IEEE J Sel Top Quantum Electron 19:95–113
74. Ogawa K, Goi K, Tan YT, Liow TY, Tu X, Fang Q, Lo GQ, Kwong DL (2011) Silicon Mach-Zehnder modulator of extinction ratio beyond 10 dB at 10.0-12.5 Gbps. Opt Express 19:B26–B31
75. Wang J, Qiu C, Li H, Ling W, Li L, Pang A, Sheng Z, Wu A, Wang X, Zou S, Gan F (2013) Optimization and demonstration of a large-bandwidth carrier-depletion silicon optical modulator. J Lightw Technol 31:4119-4125
76. Yu H, Bogaerts W (2012) An equivalent circuit model of the traveling wave electrode for carrier-depletion-based silicon optical modulators. J Lightw Technol 30:1602–1609
77. Chang WH (1976) Analytical IC metal-line capacitance formulas. IEEE Trans Microw Theory Tech 24:608–611
78. Heinrich W (1993) Quasi-TEM description of MMIC coplanar lines including conductor loss effects. IEEE Trans Microw Theory Tech 41:45–52
79. Sheng Z, Wang Z, Qiu C, Li L, Pang A, Wu A, Wang X, Zou S, Gan F (2012) A compact and low-loss MMI coupler fabricated with CMOS technology. IEEE Photonics J 4:2272–2277
80. Qiu C, Sheng Z, Li H, Liu W, Li L, Pang A, Wu A, Wang X, Zou S, Gan F (2014) Fabrication, characterization and loss analysis of silicon nanowaveguides. J Lightw Technol 32:2303–2307
81. Neamen DA (2003) Semiconductor physics and devices: basic principles. McGraw Hill
82. Dong P, Liao S, Liang H, Qian W, Wang X, Shafiiha R, Feng D, Li G, Zheng X, Krishnamoorthy AV, Asghari M (2010) High-speed and compact silicon modulator based on a racetrack resonator with a 1 V drive voltage. Opt Lett 35:3246–3248
83. Watts MR, Zortman WA, Trotter DC, Young RW, Lentine AL (2010) Low-voltage, compact, depletion-mode, silicon Mach-Zehnder modulator. IEEE J Sel Top Quantum Electron 16:159–164
84. Baehr-Jones T, Ding R, Liu Y, Ayazi A, Pinguet T, Harris NC, Streshinsky M, Lee P, Zhang Y, Lim AE, Liow TY (2012) Ultralow drive voltage silicon traveling-wave modulator. Opt Express 20:12014–12020
85. Ding J, Ji R, Zhang L, Yang L (2013) Electro-optical response analysis of a 40 Gb/s silicon Mach-Zehnder optical modulator. J Lightw Technol 31:2434–2440

Chapter 3
CMOS-Compatible Advanced Multiplexing Technology

Abstract In this chapter, we study the advanced multiplexing scheme in CMOS-compatible silicon-based optical interconnection. First, we introduce the latest progress of the three most widely used multiplexing schemes, WDM, polarization multiplexing (PDM) and MDM. Then the design theory, layout drawing and experimental validation of AWG for DWDM applications were studied. Then two novel polarization beam splitter (PBS) and MDM devices for ultra-broadband applications are studied, and their performance and processing tolerance are analyzed. The multiplexing of information is the most commonly used technology to improve the communication capacity. The typical multiplexing technology includes time, wavelength, mode, polarization and orbital angular momentum [1, 2]. In this chapter, we study three multiplexing technology for silicon optical interconnections, which are wavelength-division multiplexing (WDM), polarization-division multiplexing (PDM) and mode-division multiplexing (MDM).

3.1 WDM: Low-Loss and Low-Crosstalk 8-Channel AWG Router

3.1.1 Research Progress of Si-AWG Technology

Arrayed waveguide grating (AWG) is a very important filter and the only commercially available integrated filter. The first AWG was proposed in 1988 by Mint Smit from Delft University of Technology [3]. After more than 20 years of development, AWG has been implemented on a variety of material systems, including silicon-based silicon dioxide (SiO_2/Si) [4], silicon-based germanium (Ge/Si) [5], silicon-on-insulator (SOI) [6], InGaAsP/InP [7], polymer [8], lithium niobate ($LiNbO_3$) [9], SiON [10], Silicon nitride (SiN) [11], etc. The wavelength is also extended from the traditional optical communication band (1310, 1550 nm) to other bands such as

J. Wang, *CMOS-Compatible Key Engineering Devices for High-Speed Silicon-Based Optical Interconnections*, Springer Theses,
https://doi.org/10.1007/978-981-13-3378-1_3

mid-infrared (5 μm) [5, 12]. Among them, because of the strong refractive index difference between Si and SiO_2 ($\delta n \sim 2$), the device size is reduced by three orders of magnitude compared with the traditional silicon dioxide device. Moreover, SOI devices can be fabricated by large-scale and low-cost CMOS process, so the AWG based on silicon nanowire optical waveguide has attracted much attention in recent years.

In 2004, researchers at Yokohama National University reported a saddle-type Si-AWG with channel spacing of 6 nm and channel number of 17, which reduced the device size to 100 square microns for the first time, but the insertion loss exceeds 10 dB and the crosstalk is as high as –5 dB [6]. In 2005, the research group reported a smaller disk-type Si-AWG and improved its device performance [13]. In 2006, the device performance was further improved by optimizing the transition between different waveguides, reducing the diffraction series and reducing the length of arrayed waveguides [14].

Researchers in Ghent University reported a Si-AWG with 3 nm channel spacing in 2004. The loss was about 8 dB and the crosstalk is about –6 dB [15]. In 2006, the group introduced a double-etched taper near the free propagation region (FPR) to reduce the mode matching loss in the arrayed waveguide and the slab region. They reported a 4-channel Si-AWG router with 2 nm channel spacing, 3.5 dB loss and –12 dB crosstalk [16]. In the same year, the effect of phase error on loss and crosstalk was effectively reduced by introducing a wide arrayed waveguide [17]. A 16-channel Si-AWG with 1.6 nm channel spacing was reported. The loss and crosstalk are only 2.2 dB and –17 dB, respectively. In 2010 and 2011, the device performance were further optimized in a 8-channel demultiplexer [18] and a 16 channel router with channel spacing of 3.2 nm [19], respectively. Between 2012 and 2013, the research team adopted polymer-covered narrow arrayed waveguides [20], two-dimensional grating polarization diversity mechanism [21], and multi-mode interference couplers [22], to solve the problems of temperature sensitivity, polarization sensitivity and narrow spectrum bandwidth of traditional Si-AWG devices, respectively. These efforts greatly promotes the research of Si-AWG devices. In 2013, the group effectively increased the free spectrum range (FSR) by reducing the spacing between the arrayed waveguides while maintaining the device size. They reported three high-performance Si-AWGs with different channel spacing at one time [23]. In 2014, the team found that the layout resolution caused a deviation in the length of the Si-AWG arrayed waveguide, resulting in exceptional phase errors [24]. By using a 1-nm resolution layout, the device with 1.6 nm/3.2 nm channel spacing exhibit a loss of 2 dB/1.5 dB and a crosstalk of –22.5 dB/26 dB, respectively. It has been the best performance reported so far. In the same year, they also compared the performance differences between AWG and EDG with different channel spacings [25].

Other foreign research teams on Si-AWG and their work are also introduced briefly here. The researcher from Electronics and Telecommunications Research Institute (ETRI) used rib waveguides to reduce the effect of waveguide width errors on crosstalk in 2008 [26], and then studied the effects of process differences at different positions on the central wavelength of Si-AWG on SOI wafers in 2009 [27]. People from University of California, Davis, realized a Si-AWG with the largest number of channels (512) and the smallest channel spacing (0.2 nm) in 2012 [28, 29]. The reflective Si-AWGs implemented by AiDi Corporation, Japan have no curved waveguides, and some MMI couplers are introduced to filter out the multimode in the waveguides. So the device performance is obviously improved compared with other kinds of reflective AWG devices [30, 31].

China's research on Si-AWG is lagging behind foreign countries for several years. Compared with the researchers from Yokohama National University in Japan and Ghent University in Belgium, who are devoted to disk-type and saddle-type devices respectively, the researchers from Zhejiang University (ZJU) in China have done a lot of work in the design and implementation of novel devices. In 2006, Dai et al., proposed two kinds of ultra-small Si-AWG devices with overlapped [32] and micro-bended arrayed waveguide gratings [33], respectively. In the same year, first type of ultra-small Si-AWG device based on amorphous silicon was demonstrated experimentally [34], which has been the smallest Si-AWG, with an area of only $40 \times 50 \mu m^2$. The second kind of devices was also demonstrated in 2011 [35]. Since then, they have realized some reflective AWGs with only one FPR, and proposed new device designs of polarization insensitivity and spiral arrayed waveguide, respectively. In 2015, they used the Mach-Zehnder interferometer (MZI) as an optical interleaver and integrated it with 3.2 nm-channel-spacing AWGs to achieve 1.6 nm-channel-spacing wavelength division multiplexing [36]. Then they also combined wavelength division multiplexing and polarization multiplexing by using bi-directional AWG and polarization controllers [37]. Lang et al., also did some special work on resolving the polarization sensitivity of devices. A tilted star coupler and a new method of designing different diffraction orders for different polarization states were proposed [38]. In 2011, they reported theoretical results on birefringence compensation in Si-AWG channels [39]. In 2013, a birefringence-compensated ultra-small Si-AWG triplexer was achieved with only $180 \times 120 \mu m^2$ footprint, 3~4 dB loss, and less than –18 dB crosstalk. The polarization-dependent wavelength shift is reduced to 2.5 nm [40]. In 2014, a 5-channel birefringence-compensated Si-AWG was implemented with a channel spacing of 20 nm for CWDM applications [41]. The device size was $360 \times 170 \mu m^2$, and the polarization-dependent wavelength shift was reduced to 0.5~3.5 nm. In addition to Zhejiang University, the Institute of Semiconductors (IS-CAS) [42–45], Huazhong University of Science and Technology [46] and Tianjin Polytechnic University (TJPU) [47–51] have also done some meaningful research.

In addition to the basic applications as wavelength multiplexing, demultiplexing, routing and triplexing, the reported Si-AWG also includes the application of gas sensors which was realized by cascading micro-ring resonators and Si-AWG [52], and the verification of spectrometers in mid-infrared band [12]. In 2007, the researchers implemented a 50-channel Si-AWG spectrometer with 0.2 nm channel spacing [53]. The device was used to demodulate in the fiber Bragg grating refractometer [54].

Table 3.1 summarized the experimental results of Si-AWG for wavelength multiplexing, demultiplexing, routing, and triplexing.

Table 3.1 Experimental summary of the previous Si-AWG (2004–2015.5)

Group	Time	A(μm^2)	$\Delta\lambda_{CH}$(nm)	N_{CH}	FSR(nm)	IL(dB)	L_u(dB)	XT(dB)	References
YNU	2004	110×93	6	17	90	10	2	–5	[6]
YNU	2005	70×60	11	9	92	1	2.5	–13	[13]
YNU	2006	70×75	8	11	140	2	2.5	–13	[14]
Ghent	2004	380×290	3	8	24	8	3	–6	[15]
Ghent	2006	425×125	2	4	8	3.5	5	–12	[16]
Ghent	2006	500×200	1.6	16	25.6	2.2	3	–17	[17]
Ghent	2010	200×350	3.2	8	*	1.1	1.3	–25	[18]
Ghent	2011	475×330	3.2	16	51.2	3	2.09	–19	[19]
Ghent	2012	350×250	3.2	8	*	2.6	0.7	–15	[20]
Ghent	2012	1400×850	1.6	16	*	2.6	4	–21.5	[21]
Ghent	2013	560×350	3.2	12	*	3.29	1.55	–17	[22]
Ghent	2013	1180×285	0.8	4	6.9	2.45	0.29	–17	[23]
Ghent	2013	540×320	2	8	24.8	1.32	0.6	–19	[23]
Ghent	2013	380×330	3.2	12	29	0.53	0.6	–23	[23]
Ghent	2014	920×446	1.6	16	29	2	1.7	–22.5	[24]
Ghent	2014	530×435	3.2	16	54	1.5	2	–26	[24]
Ghent	2014	275×245	6.4	4	38	1.53	0.37	–27.1	[25]
Ghent	2014	310×305	6.4	8	64	1.86	1.16	–27.3	[25]
Ghent	2014	305×260	24	4	88	1.68	1.10	–20.6	[25]
Ghent	2014	305×260	32	4	144	2.1	1.75	–19.1	[25]
ZJU	2006	40×50	11	4	*	10	2	–10	[34]
ZJU	2010	134×115	3.2	8	28	3	3	–12	[55]
ZJU	2010	193×168	1.6	16	22	*	*	*	[56]
ZJU	2011	163×147	3.2	15	*	7.5	5	–8	[35]
ZJU	2015	520×190	1.6	18	*	8	2	–18	[36]
ZJU	2013	180×120	*	*	*	3	*	–18	[40]
ZJU	2014	360×170	20	5	*	3	*	–18	[41]
AiDi	2012	*	0.8	40	*	3	5	–15	[28]
AiDi	2013	530×230	3.2	14	*	3	2.5	–20	[30]
AiDi	2014	1600×1100	0.2	512	*	*	*	–4	[29]
ETRI	2008	*	1.6	16	25.8	3	3	–16	[26]
ETRI	2008	*	3.2	8	25.8	3	3	–18	[26]
IS-CAS	2011	70×50	*	*	*	8	8	–15	[43]
IS-CAS	2011	580×170	1.6	16	25.6	2.2	2.5	–8	[44]
IS-CAS	2014	*	3.2	8	25.6	2.4	4	–25.1	[42]
TJPU	2013	230×160	2	8	*	3.18	1.35	–23.1	[49]

*denotes no performance characterization from the reference in the table

3.1.2 Operation Principle of Si-AWG

Figure 3.1 shows a schematic of $N_{in} \times N_{out}$ AWG, which consists of N_{in} input waveguides, N_{out} output waveguides, two free propagation regions (FPR) and a number of arrayed waveguides. These waveguides are uniformly arranged around FPR with certain gaps. The operation principle of AWG can be described as follows: when a light beam enters into the input FPR, diffraction will occur, and the expanded diffracted beam will be coupled into the arrayed waveguides. Because of the fixed optical length difference between the adjacent arrayed waveguides, light at different wavelengths will have different phase differences. So the constructive or destructive interference will occur in different output waveguides. The function of wavelength multiplexing and de-multiplexing can be realized.

The detailed derivation of the operation principle of AWG can be referred to the doctoral thesis of Daoxin Dai from Zhejiang University, Huamao Huang from Huazhong University of Science and Technology, Pieter Dumon/Shibnath Pathak from Ghent University [57–60]. Most of the derivations are omitted in this article, and only some important results are given.

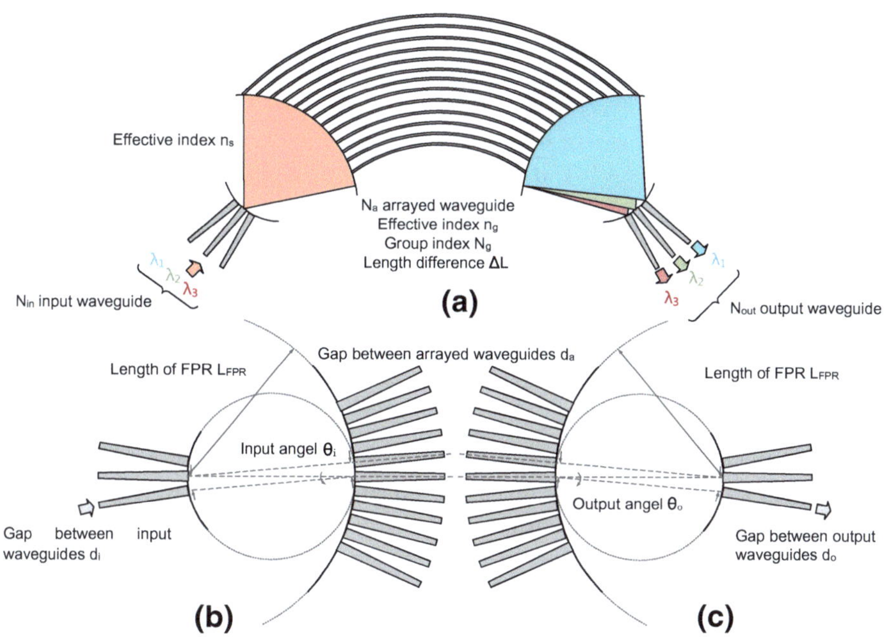

Fig. 3.1 **a** Schematic of AWG with $N_{in} \times N_{out}$ channels. **b** Local schematic of input FPR. **c** Local schematic of input FPR. AWG paramters: N_{in} input waveguide, N_{out} output waveguide, N_a arrayed waveguide, length difference between arrayed waveguides ΔL, effective refractive index of arrayed waveguide n_g, group index N_g, FPR length L_{FPR}, gap between arrayed waveguides d_a and gap between input/output waveguides $d_i = d_o = d$

In this thesis, only the symmetric AWG is considered. Suppose that light at wavelength of λ inputs from the input waveguide with input angle θ_i and outputs from the output waveguide with output angle θ_o. If we want constructive interference at the output waveguide, the total optical path difference of the two beams of the adjacent arrayed waveguides should be a multiple of the wavelength λ(m is the diffraction order). Therefore, the grating equation of AWG can be obtained.

$$n_g(\lambda)\Delta L - n_s(\lambda)d_a sin\theta_o - n_s(\lambda)d_a sin\theta_i = m\lambda \tag{3.1}$$

In the above derivation, considering $L_{\text{FSR}} \gg d_a$ generally, $\Delta L^{in} \approx d_a sin\theta_i$ and $\Delta L^{out} \approx d_a sin\theta_o$, are approximately used to calculate the optical path difference in the free propagation region. The non-approximate formula and the error analysis can be seen in the Ref. [57]. For the central wavelength λ_o, when light inputs from the central input waveguide and outputs in the central output waveguide, $\theta_i = \theta_o = 0$, thus Eq. 3.1 can be simplified to:

$$n_g(\lambda_0)\Delta L = m\lambda_0 \tag{3.2}$$

For non-central wavelengths, in order to investigate the light focusing position on the output facet, we assume that the input position is fixed. The angle dispersion equation of the output waveguide can be obtained by deriving the wavelength λ and neglecting the small term on both sides of Eq. 3.1.

$$\frac{d\theta_o}{d\lambda} = \frac{mN_g(\lambda)}{n_s(\lambda)n_g(\lambda)d_a} \tag{3.3}$$

Here, $N_g(\lambda) = n_g(\lambda_0) - \lambda_0 \frac{d}{d\lambda}[n_g(\lambda)]$ is the group refractive index in the arrayed waveguide. Assume that the channel spacing of AWG is $\Delta\lambda_{\text{CH}}$, then the gap d between adjacent output waveguides can be obtained according to Eq. 3.3:

$$d = \frac{m\Delta L N_g(\lambda)L_{\text{FPR}}}{n_g(\lambda_0)n_s(\lambda)d_a} \tag{3.4}$$

Due to the periodicity of the grating imaging, if the m-order imaging position at wavelength λ overlaps with the m-1 order imaging position at wavelength λ+FSR, such FSR is called the free spectrum range.

$$\text{FSR} = \frac{\lambda n_g(\lambda)}{\text{mN}_\text{g}(\lambda)} = \frac{\lambda}{\text{M} - 1} \tag{3.5}$$

Here M is modified diffraction order. FSR is a very important index, which defines the maximum number of operation channels $N_{\max}$.

$$N_{\max} = int(\frac{\text{FSR}}{\Delta\lambda_{\text{CH}}}) \tag{3.6}$$

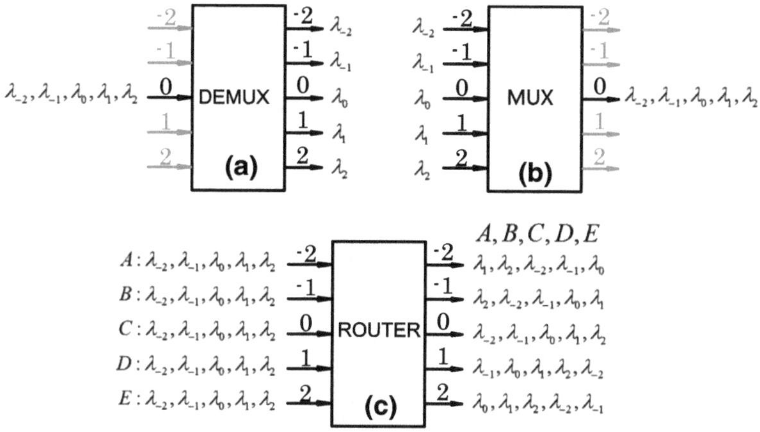

Fig. 3.2 Schematic of AWG's functions, where $\lambda_i = \lambda_0 + i\Delta\lambda_{CH}$. Order number refers to Fig. 3.1a

When $N_{in} = N_{out} = N_{max} = \frac{FSR}{\Delta\lambda_{CH}}$, the AWG works as a router device. When AWG is symmetric and can input from the i-th port, the wavelength in the j-th port must be $\lambda + (i + j)\Delta\lambda_{CH}$. Figure 3.2 shows some AWGs working for wavelength multiplexing, de-multiplexing and routing.

3.1.3 Layout Preparing of Si-AWG

After determining the basic parameters of the device, it is necessary to select the appropriate layout structure to design the device completely. In addition to the mode converter near FPR, the traditional disk-type AWG layout waveguide are all curved, which is only suitable for the case of small diffraction order (large channel spacing). When the diffraction order is high (small channel spacing), the distance between the arrayed waveguides will be much larger than the de-coupling distance (~2 μm) to achieve larger arrayrf waveguide length difference. Thus the device becomes very large. The FPR-overlapped or reflective AWG layout, although the device size is small, has produced additional performance degradation. Therefore, in this thesis, we choose a saddle-shaped AWG layout to achieve a compromise size and high-performance.

The AWG layout is generated by using Matlab programming in Clewin software. Figure 3.3 shows a schematic of saddle-shaped AWG layout with mixed wide and narrow waveguides and a double-etching mode converter near the free propagation region. In this work, we only consider the case of 90 degree angle. This AWG contains three main parts:

Fig. 3.3 Schematic of AWG layout

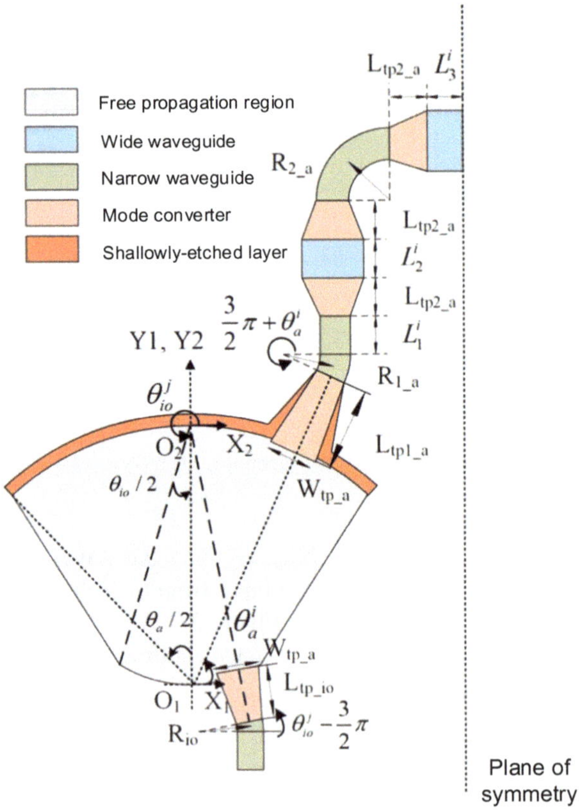

1. FPR: It is formed by closing two arcs with radii L_{FPR} and $L_{FPR}/2$ respectively.
2. Input and output waveguide: It consists of a mode converter with a length of L_{tp_io}, a curved waveguide with a radius of R_{io}, and a straight waveguide connecting to external structures.
3. Arrayed waveguide: It includes a double-etching mode converter with a length of L_{tp1_a}, a curved waveguide with a radius of R_{1_a}, a narrow straight waveguide with a length of L_1^i, 3 mode converters with a length of L_{tp2_a} between the broad and narrow waveguides, a 90 degree narrow curved waveguide, and two wide straight waveguides with lengths of $L_2^i L_3^i$. In the arrayed waveguide region, 4 parts need to be adjusted according to the number i of the arrayed waveguide, in which the narrow straight waveguides are used to offset the optical path difference of the curved waveguide with several angles. Therefore, the optical path difference of the arrayed waveguide is only related to two segments of the straight waveguides, which simplifies the design.

Table 3.2 Basic parameters for drawing a AWG layout

Parameter	Symbol
The coordinates of O_1 in the coordinate system {O2; X2, Y2}	(x0, y0)
Width of wide waveguide	W_w
Width of narrow waveguide	W_n
Length of free propagation region	L_{FPR}
Number of input/output waveguide	N_{CH}
Gap between input/output waveguide	d
Length of tapered mode converter	L_{tp_io}
Waveguide width of tapered mode converter near FPR	W_{tp_io}
Bend radius in the input/output waveguides	R_{io}
Number of arrayed waveguide	N_a
Gap between arrayed waveguides	d_a
Length of double-etched tapered mode converter	L_{tp1_a}
Waveguide width of double-etched tapered mode converter near FPR	W_{tp2_a}
Length of tapered mode converter	L_{tp2_a}
Bend radius in the arrayed waveguides	R_{1_a},R_{2_a}
Length difference between arrayed waveguides	ΔL
Lengths of straight wide waveguides in the shortest arrayed waveguide	L_2^0, L_3^0

The AWG layout can be drawn as follows:

1. First, determine the basic parameters in Table 3.2.
2. The free propagation region is determined by position coordinates (x0, y0) of AWG, FPR length L_{FPR} and the largest radian angles θ_{io} and θ_a.
3. The layout of the input and output waveguides is drawn in the coordinate system {O2; X2, Y2}. First, the coordinates between the waveguide and FPR are determined by the number N_{CH} of the input/output waveguides and the waveguide gap d. It is important to note that there is an endpoint at the starting end of the tapered mode converter which can not be on the interface (unlike the interface of the arrayed waveguide). Therefore, we need to pay more attention to drawing, and we can take several attempts to ensure that this point is within FPR.
4. The layout of the arrayed waveguide area is completed in the coordinate system {O1; X1, Y1}. First, the coordinates of the arrayed waveguides at the interface are determined by the number N_a of the arrayed waveguides and the waveguide gap d_a. There are two points to be noticed in the layout of this area: one is the length of the narrow straight waveguide L_1^i. The radius of the bend waveguide used in this work is large enough to approximate that the effective refractive index of the mode is the same as that of the straight waveguide. Therefore, the length difference of the narrow waveguide can be obtained according to the optical path difference of the bent waveguides of the adjacent arrayed waveguides. The second is the lengths of the wide straight waveguides, L_2^i and L_3^i. L_3^i can be

obtained according to the conditions of the symmetry plane, and then L_2^i of the other waveguides is obtained from the shortest arrayed waveguide according to the iterative relationship of the length difference $\Delta L = L_3^{i+1} - L_3^i + L_2^{i+1} - L_2^i$.
5. After determining the layout of the left half side, you can use the mapping function of Clewin software to copy the layout on the right. In addition, in the nearby region of FPR, the accuracy of the process should be considered, and the gap between adjacent tapered waveguides should be set reasonably.

3.1.4 8×3.2 nm Si AWG Router Based on Different Optimization Technologies

For Si AWGs, a very important factor that restricts its performance is the surface roughness of the arrayed waveguide. Because the effective refractive index of the mode is related to the shape of the cross-section of the waveguide, the surface roughness of the waveguide will cause the jitter of the effective refractive index of the mode, and then influence the parameter ΔL of the AWG. This will not only cause a channel spacing difference $\Delta\lambda_{CH}$, but also cause the deterioration of crosstalk and loss performance. For the saddle-shaped AWG, the methods of reducing the influence of the surface roughness on the crosstalk performance include using a ridge waveguide instead of the strip waveguide [26] and using wide arrayed waveguide instead of the narrow arrayed waveguide [18], etc.

Another problem to consider is the mode conversion between different types of waveguides. For AWGs, the modes in the FPR, input/output waveguide and arrayed waveguide are different. Especially there is very obvious mode mismatching loss between the slab mode in the FPR and the guided modes in the other two kinds of waveguides. Therefore, it is necessary to use suitable mode converters among different regions. A common mode converter is a tapered mode converter. By slowly increasing the width of the waveguide, the width of the waveguide near FPR is large enough to reduce the mode mismatching loss. You can also take a parabolic or sinusoidal shaped mode converter, which may reduce the device length [42]. In addition, a double etching tapered mode converter can be used to slowly convert the strip waveguide to ridge waveguide. Because the modes in ridge waveguide and slab waveguide is more similar, it will also reduce the loss [18].

According to the technical conditions provided by the process line, we adopted two optimization techniques for AWG.

1. 1 μm wide transmission waveguides are used in the arrayed waveguide region. The waveguide with 0.45 μm width is still used to limit the light, and thus reduce the radius of the bent waveguide (~5 μm). Narrow waveguides and wide waveguides are connected by tapered mode converters.
2. The width of the waveguide is increased from 0.45 μm to 2.5 μm by using a 70 nm/220 nm double etching mode converter with a length of 20 μm at the edge of the FPR and arrayed waveguide.

Table 3.3 Design parameters of AWG with channel spacing of 3.2 nm

	♯1	♯2	♯3
Channel spacing ($\Delta\lambda_{CH}$) (nm)	3.2	3.2	3.2
Channel number (N_{CH})	8	8	8
Free spectrum range (FSR) (μm)	25.6	25.6	25.6
Rowland radius (L) (μm)	130.967	131.725	131.725
Arrayed waveguide number (N)	35	35	35
Length difference of arrayed waveguide (ΔL) (μm)	23.088	25.4	25.4
Gap between input/output waveguides (d)(μm)	3.2	3.2	3.2
Gap between arrayed waveguides near FPR (d_a) (μm)	2.8	2.8	2.8

We designed some 8-channel Gaussian AWG devices with channel spacing of 8 nm, 3.2 nm, 1.6 nm and 0.8 nm, respectively. And we also tried to design a 8-channel flat-top AWG device with channel spacing of 3.2 nm. In this section, we mainly discuss 8-channel Gaussian AWG devices with 3.2 nm channel spacing, and study the influence of two optimization techniques on their performance. We have designed three AWG devices (♯1, ♯2, ♯3). AWG (♯1) does not use any optimization technology, AWG (♯2) uses wide waveguides in the arrayed waveguide region, and AWG (♯3) uses both of the two optimization techniques. The design parameters of these three devices are shown in Table 3.3.

These devices are processed in the silicon photonic fabrication platform from Shanghai based on 130 nm standard CMOS process. SOI wafer with 220 nm top silicon layer and 2 μm BOX layer are used to prevent the mode from leaking into the substrate and causing additional transmission loss. The minimum waveguide size is 150 nm and the minimum gap is 200 nm by using 248 nm lithography technology. In order to realize strip waveguide and ridge waveguide at the same time, the fabrication platform decomposes the 220 nm silicon etching into a two-step etching process. At the same time, in order to improve the sidewall roughness of the waveguide, a SiN layer is used as the etching mask.

1. A SiN layer is grown on the SOI wafer. The SiN layer acts as a mask layer in the following etching step. As the stain coefficient between the SiN layer and the Si wafer does not match, in order to alleviate the stress, a thin layer of SiO_2 is usually grown on the Si wafer as a buffer layer before the growth of the SiN layer.
2. The Si waveguide region is defined on wafer by using 248 nm deep ultra-violet lithography.
3. The SiN mask is removed by inductively coupled plasma reactive ion etching(ICP-RIE. The etching process uses photoresist as an etching mask.
4. The second lithography is used to realize the ridge and strip waveguide structure. The photoresist covers the SiN mask in the ridge waveguide, and is removed by the development method in the strip waveguide.
5. First silicon etching.
6. The etching photoresist is removed without photolithography.

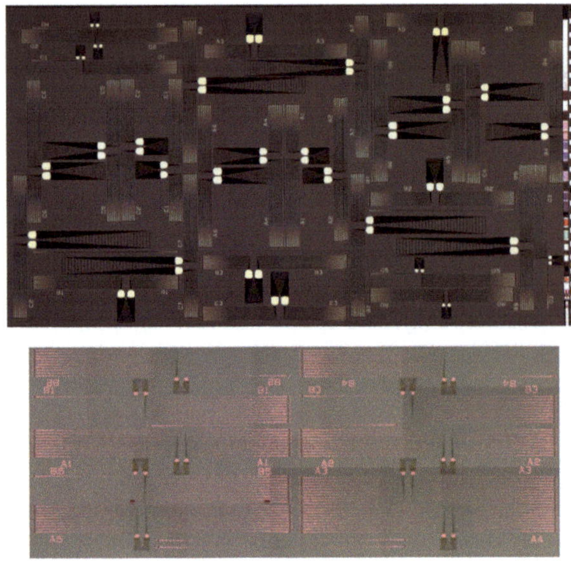

Fig. 3.4 Microscoped pictures of two AWG chips

7. Second silicon etching. After this etching process, the strip waveguide and ridge waveguide are realized simultaneously.
8. After removing the SiN mask layer through wet etching, the basic structure of the optical waveguide is realized.

Figure 3.4 shows all the AWG devices in this process. We placed AWG devices as compact as possible to reduce the layout area.

Figure 3.5a–c show the microscope picture of AWG(♯1, ♯2, ♯3), respectively. We also describe the local enlarged drawing of arrayed waveguide and mode converter near FSR.

These AWGs were designed to work at TE_0 mode, so we designed TE grating couplers on each input and output waveguides. The grating coupler are achieved by using the two-step etching process. The period is 0.63 μm, the number is 15, and the duty cycle is 50%. Figure 3.6a shows the schematic diagram of the grating coupler and its assisted waveguide. Figure 3.6b shows the microscope diagram of the processed grating coupler. In order to measure the transmission spectrum of the device, we used the a tunable laser (Agilent 8164B, 81600-201) as the light source, and used a tricyclic polarization controller to control the polarization state, and coupled the light into the central input channel of the AWG. The transmitted light is coupled to the fiber through a grating coupler and it is finally detected in the powermeter (Agilent 81624B). The actual test platform with grating coupling is shown in Fig. 3.6c.

The normalized transmission spectrum of AWG (♯1, ♯2, ♯3) are shown in Fig. 3.7a–c, respectively. To facilitate performance comparison, we extracted the insertion loss and crosstalk in each channel, as shown in Fig. 3.8a–b, respectively. Because of the serious phase errors caused by narrow arrayed waveguides, AWG

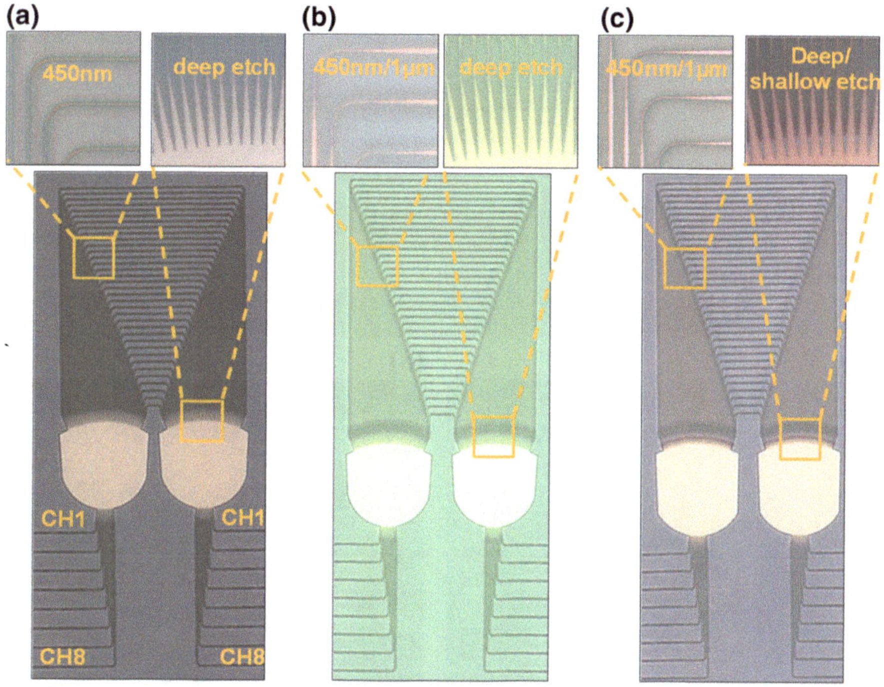

Fig. 3.5 **a–c** Microscoped picture of AWG(♯1, ♯2, ♯3), respectively. Reproduced from Ref. [61] by permission of Optical Society of America

(♯1) exhibits poor performance in insertion loss and crosstalk. The sidewall roughness of the waveguide causes the dithering of the effective refractive index of the mode, thus destroying the image quality at the output port, so that a considerable amount of optical power enters the non-designed channel, and the optical power in the designed channel will be reduced. It eventually leads to poor loss and crosstalk performance. More discussion about insertion loss, crosstalk and phase error can be referred to Ref. [26]. By broadening the arrayed waveguide to 1 μm wide, it is obvious that the performance in AWG (♯2, ♯3) has been greatly improved, because when the waveguide becomes wider, the sensitivity to the size change is reduced. This can be seen from the simulation curves of the effective refractive index to the width of the waveguide. With the increase of the width of the waveguide, the slope of the effective refractive index about the width is close to 0. These two AWGs exhibit insertion loss less than 3 dB and crosstalk less than –17 dB. Compared with AWG (♯2), AWG (♯3) achieves a further performance optimization by reducing the mode mismatching loss through shallowly etching near FPR. According to the above comparison, we find that the effect of the two optimization techniques is better than the single optimization. AWG (♯3) shows the best performance with 2.32 dB insertion loss and –20.5 dB ~ –24.5 dB crosstalk.

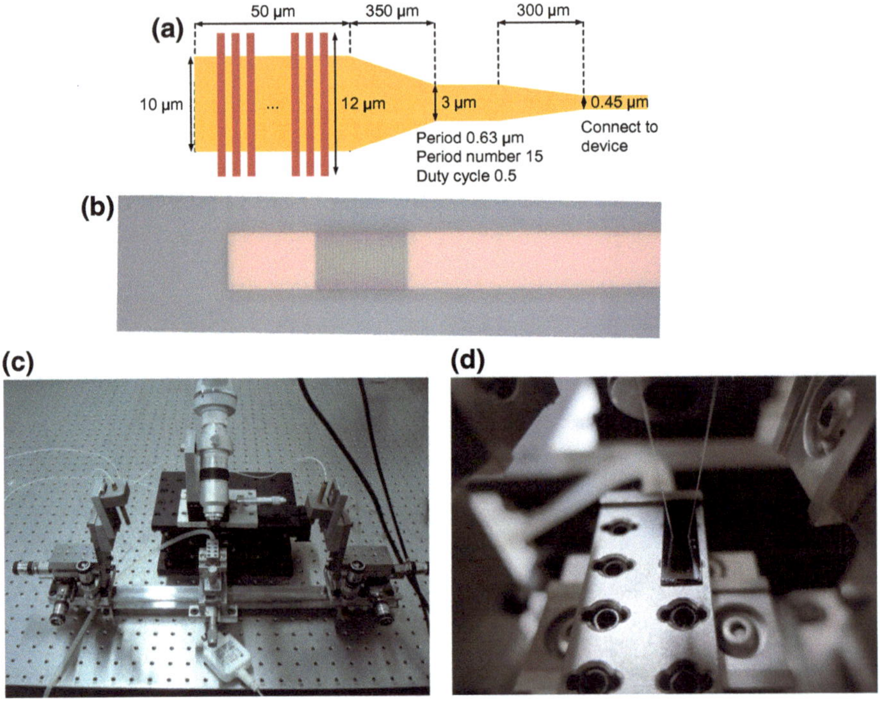

Fig. 3.6 **a** Schematic diagram of the grating coupler and its assisted waveguide. **b** Microscope diagram of the processed grating coupler. **c**–**d** Actual test platform with grating coupling

3.1.5 8×0.8 nm Si AWG Router for DWDM Applications

By using these two optimization techniques comprehensively and stable 130 nm commercial CMOS process, we have further implemented a 8 channel Si-AWG router (♯4) with 0.8 nm channel spacing for DWDM applications. The design parameters are as follows: $\Delta\lambda_{CH} = 0.8$ nm, $N_{CH} = 8$, FSR = 6.4 nm, L = 127.608 μm, N = 35, $\Delta L = 101.599$μm, m = 180, d = 3.1μm and $d_a = 2.8$μm. Figure 3.9a shows the microscope diagram of this AWG sample, which is about 4 times the size of AWG with 3.2 nm channel spacing, because the length difference of the arrayed waveguide is inversely proportional to the channel spacing. In addition, the AWG with smaller channel spacing requires not only a more comprehensive optimization method to improve the phase error, but also a good controlling of process uniformity. Figure 3.9b shows the normalized transmission spectrum at the central channel input. Figure 3.9c shows the extracted insertion loss and crosstalk performance in each channel. AWG (♯4) shows good performance with 2.92 dB insertion loss and –16.9 dB ~ –17.8 dB crosstalk. We believe that such devices with small channel spacing have great potential in silicon-based optical interconnection for DWDM applications.

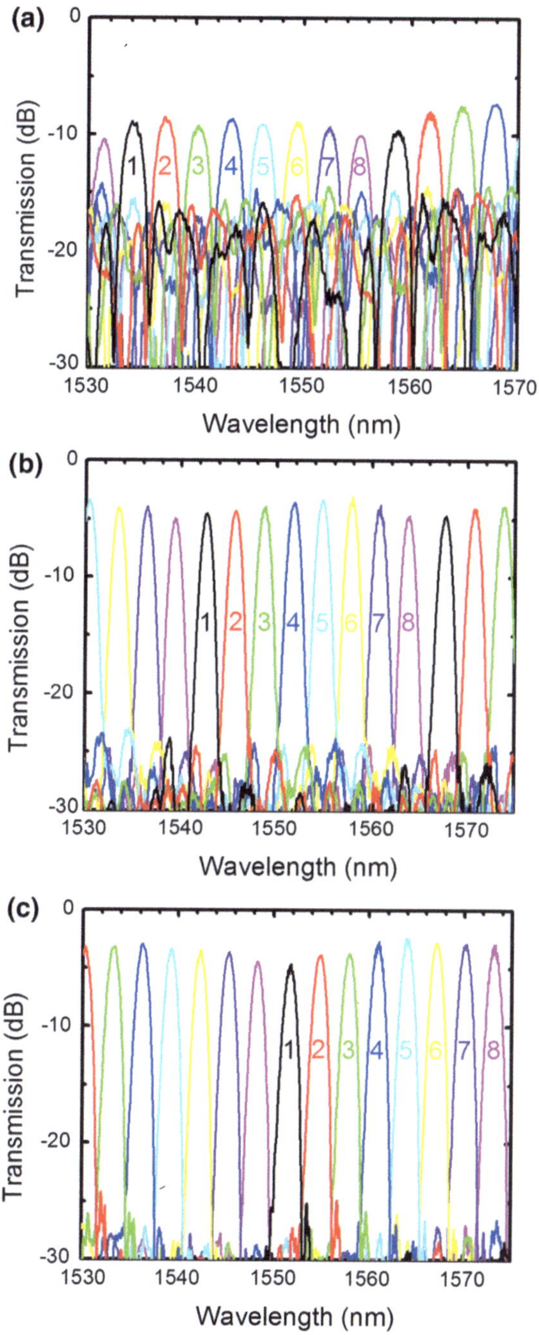

Fig. 3.7 a–c Normalized transmission spectrum of AWG (♯1, ♯2, ♯3). Reproduced from Ref. [61] by permission of Optical Society of America

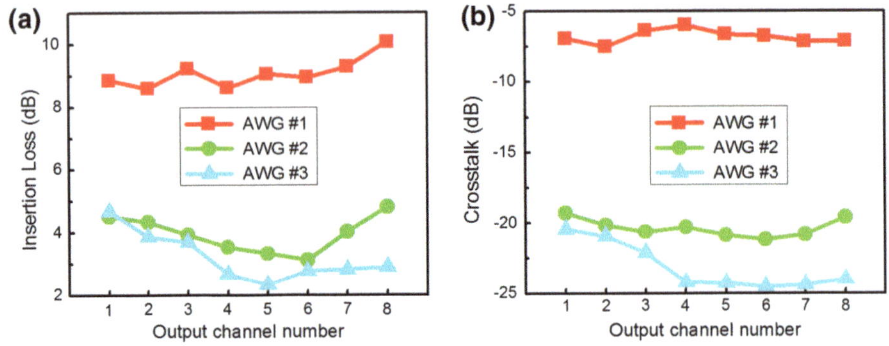

Fig. 3.8 **a–b** Insertion loss and crosstalk in each channel of AWG (♯1, ♯2, ♯3), respectively. Reproduced from Ref. [61] by permission of Optical Society of America

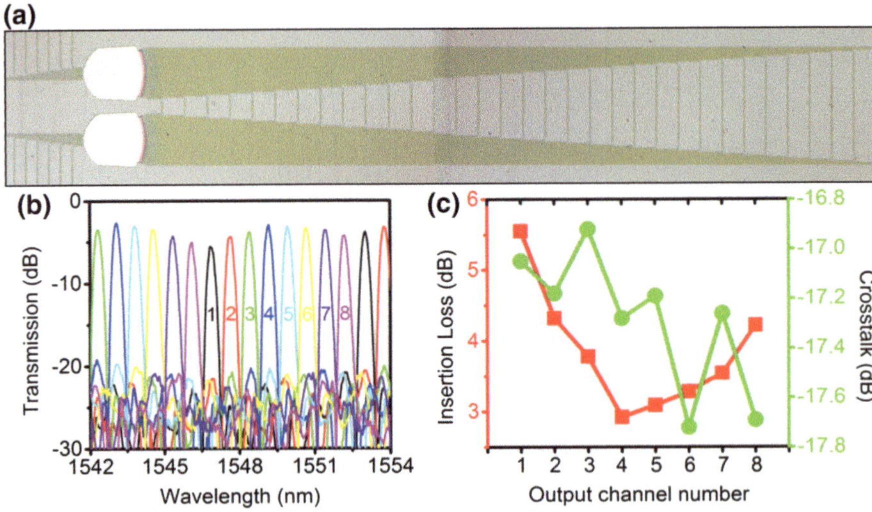

Fig. 3.9 **a** Microscope diagram of AWG (♯4) sample with 0.8 nm channel spacing. **b** Normalized transmission spectrum at the central channel input. **c** Extracted insertion loss and crosstalk performance in each channel. Reproduced from Ref. [61] by permission of Optical Society of America

3.1.6 Performance Characterization for the Cyclic Rotation Properties of the Si AWG Routers and Multi-wavelength High-Speed System Demonstration

As shown in Fig. 3.2c, the AWG router can show cyclic properties, which is an important feature of the AWG router which is different from the AWG multiplexer/demultiplexer. Figure 3.10a–b show the test results of the cyclic prosperities of AWG (♯3, ♯4), respectively. Each point in the graph represents the peak

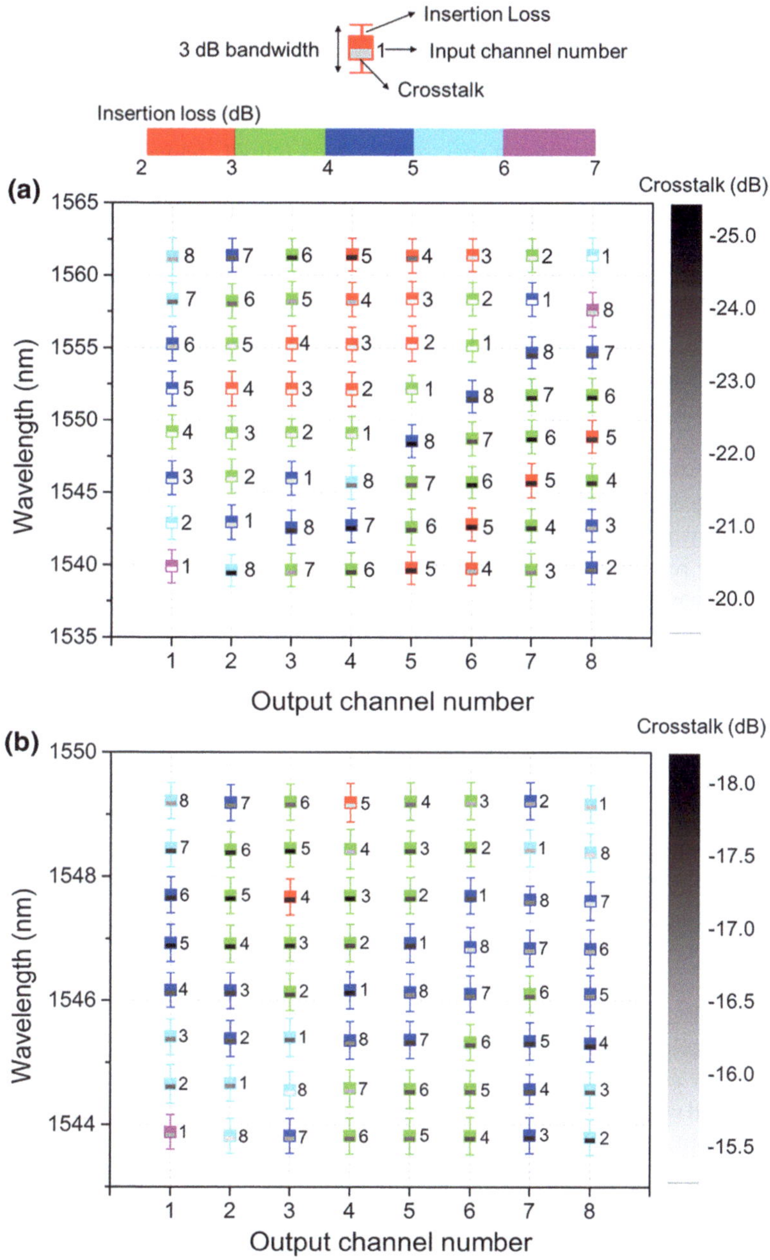

Fig. 3.10 **a–b** Cyclic rotation test results of AWG (♯3, ♯4), respectively. Reproduced from Ref. [61] by permission of Optical Society of America

wavelength of the channel, and the length of the error line represents 3 dB bandwidth, while the insertion loss, crosstalk, and the input channel mark are also marked on each point.

We find that when the input light is transferred from one channel to the adjacent input channel, the transmission spectrum of the 8 output channels will also shift by one wavelength. All the points in the figure form an aligned 8×8 matrix, showing good cyclic rotation characteristics. For AWG (♯3) and AWG (♯4), the 3 dB bandwidth are ~1.1 nm and ~0.3 nm respectively. The insertion loss is 2 dB~4 dB. And the non-uniformity are 4.13 dB and 3.66 dB, respectively. The loss non-uniformity among different channels is mainly caused by the envelope of the quasi Gauss shape in the far field, which can be improved by further increasing the length of the FPR and introducing some special structures [62]. Moreover, when the light inputs from the non-central channel, the crosstalk performance will be slightly worse. But for AWG (♯3) and AWG (♯4), the crosstalk are still less than –20 dB and –15.5 dB, respectively. Other properties include center wavelength shift and channel spacing error. The central wavelength and channel spacing are determined by the effective refractive index of the waveguide, and the two methods we used here can not accurately control the effective refractive index of the waveguide, so AWG (♯3) and AWG (♯4) do not get any improvement on center channel shift and channel spacing error. The central channel error is several nanometers, and the channel spacing error is also ~5%. Some post-fabrication tuning technology (e.g. thermal tuning) can be further used for wavelength adjustment [63].

We further study the transmission characteristics of AWG (♯3) and AWG (♯4) in a three-channel WDM system. Figure 3.11 shows the schematic of the experimental link for evaluating the high-speed transmission characteristics. We used a tunable laser (Agilent 81600-201, TLS 3) to control the wavelength of the main channel (Channel 4) signal, and then used two lasers (TLS 2, Agilent 81940A; TLS 3, Agilent 81960A) as the light source of the adjacent two channels (Channel 3 and Channel 5). The high-speed signal generator (PPG) is implemented in combination with a 10 Gb/s error-bit-ratio tester (Centellax TG1B1A) and a clock source (Hittite HMC-T2100) to produce a PRBS electrical signal of the 2.5 Gb/s, 5 Gb/s and 10 Gb/s with 2^{31}–1 length, respectively. The modulation signals of channel 3 and channel 5 are decorrelated by a 1 km long standard single mode fiber (SSMF), and then amplified by an erbium doped fiber amplifier (EDFA, PriTel SPFA-18-600). Then, the amplified mixed signals will go through a standard single mode fiber (SSMF) with a length (0 km, 5 km, and 15 km) into the central channel of AWG (Channel 4). In this experiment, AWG is used as a demultiplexer to de-multiplex the main signal in the mixed signal to the fourth channels of the output. The signal from the fourth channels is amplified by a EDFA and is controlled by an variable optical attenuator (VOA). Finally, the signal is divided into two paths through a 3 dB coupler. One signal passes through a 10 GHz receivers (PD-TIA, Agere R2860D) with the 10 Gb/s error-bit-ratio tester(BERT, Centellax TG1B1A) to detect the bit error ratio. Another signal enters into a high-speed sampling oscilloscope (DCA, Agilent 86100D and 86116C-040) to observe the eye diagram. The actual test environment is shown in Fig. 3.12.

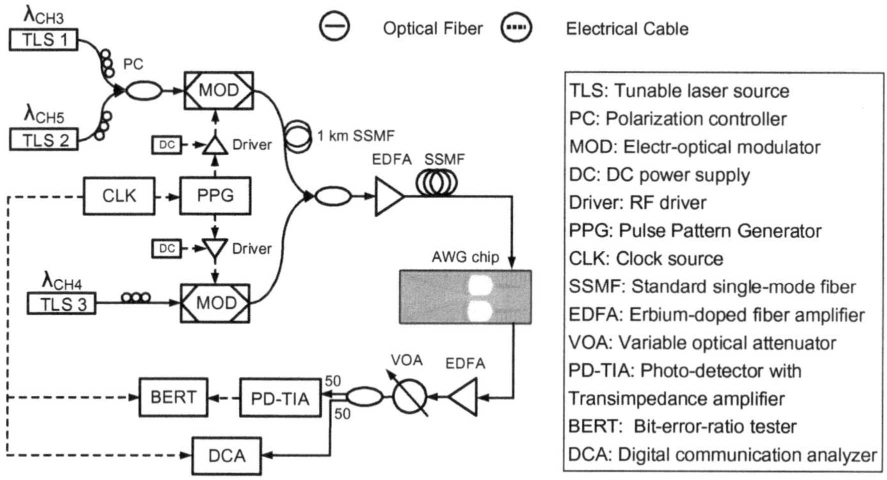

Fig. 3.11 Schematic of the experimental link for evaluating the high-speed transmission characteristics of AWGs (♯3, ♯4). Reproduced from Ref. [61] by permission of Optical Society of America

Figure 3.13a shows the test results of BER in channel 4 after demultiplexing. We also tested the back-to-back (B2B) bit-error-ratio without AWG as a reference. For the same signal speed, the power penalty of the link containing AWG (♯3) is smaller. This is because AWG (♯3) has smaller intra-channel crosstalk. Moreover, because the extinction ratio of the signal is also affected by crosstalk, the power penalty is better than the 0 km link containing AWG (♯4), even if the signal passes through a 15 km optical fiber for a link containing AWG (♯3). Nevertheless, for a 10 Gb/s 15 km long link containing AWG (♯4), less than 1.2 dB power penalty can be obtained. Because these two AWGs have smaller crosstalk in the channel, the optical eye diagram of the output signal is very clear. 10 Gb/s eye diagram under different fiber lengths is shown in Fig. 3.13b.

3.1.7 *Si-AWG Router with Other Channel Spacing and Flat-Top Spectrum*

We also implemented a Si-AWG router with 1.6 nm channel spacing in this process. The design parameters are as follows: $\Delta\lambda_{CH} = 1.6$ nm, $N_{CH} = 8$, FSR = 12.8 nm, L = 115.259μm, N = 35, $\Delta L = 50.799$ μm, d = 2.8 μm, and $d_a = 2.8$ μm. Figure 3.14a–b show the microscope picture and transmission spectrum of this AWG sample, respectively. We found that the insertion loss is 2.90 dB, the channel uniformity is 2.60 dB, and the crosstalk is about –17~–19.3 dB.

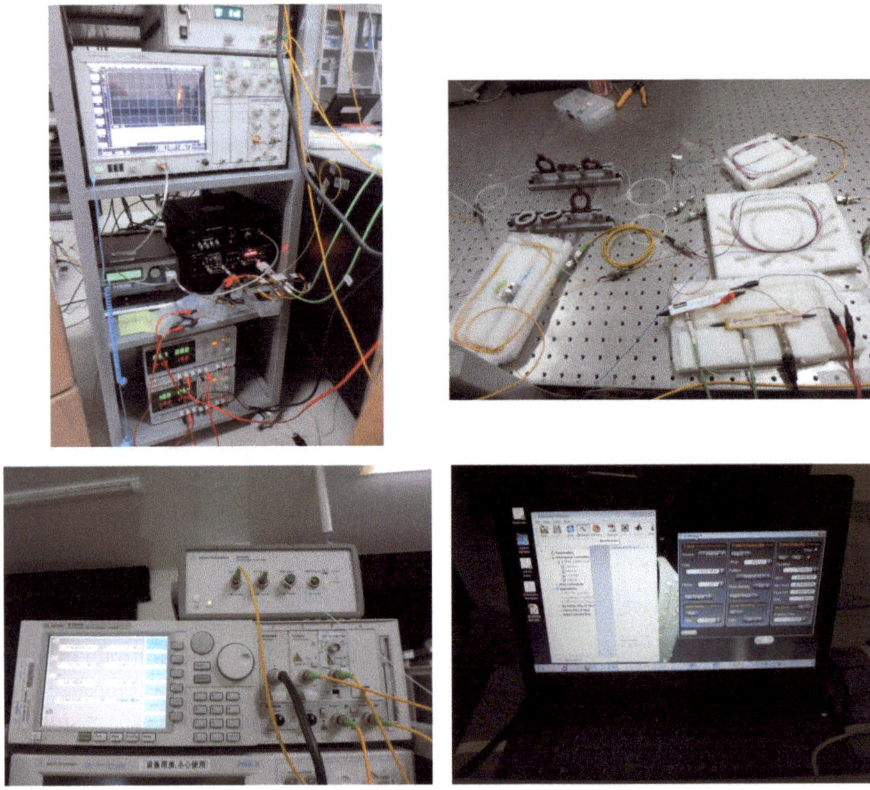

Fig. 3.12 Actual test environment for evaluating high-speed transmission characteristics of AWG (♯3, ♯4)

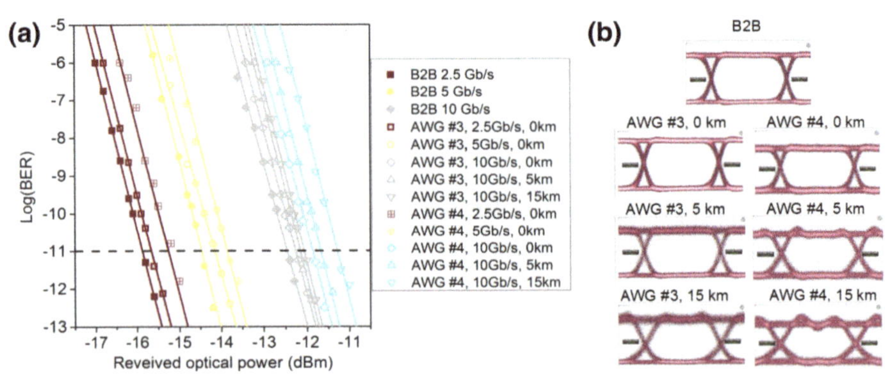

Fig. 3.13 **a** BER measurement for the demultiplexed output signal when the input signals are modulated at various data rate (2.5 Gb/s, 5 Gb/s and 10 Gb/s) and passed through various lengths (0 km, 5 km and 15 km). **b** Optical eye diagrams of the output signal modulated at 10 Gb/s data rate for fibers with various lengths. Reproduced from Ref. [61] by permission of Optical Society of America

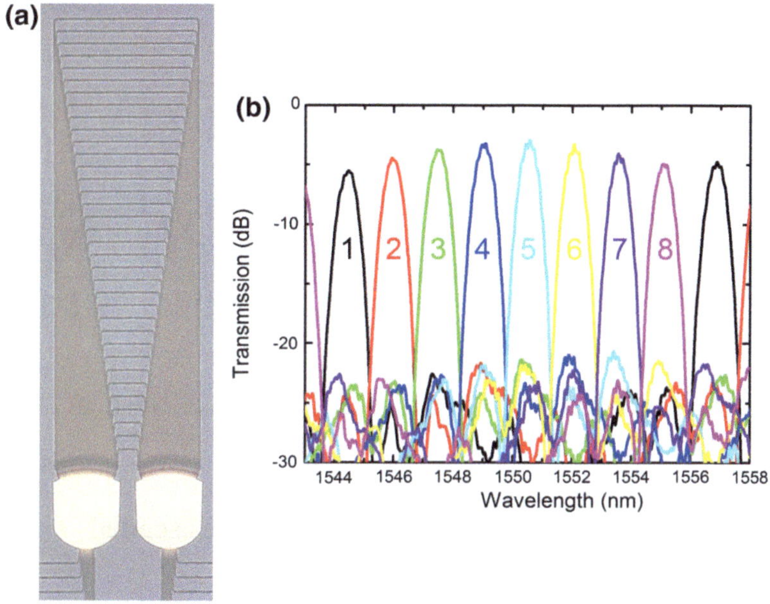

Fig. 3.14 **a** Microscoped picture of Si-AWG router sample with 1.6 nm channel spacing. **b** Normalized transmission spectrum of this sample

In the previous chapter, we reported some Gaussian Si-AWG. The 3 dB bandwidth of this kind of AWG was small, and the external factors (such as temperature drift, etc.) would affect the performance of the device. Therefore, we learned from the traditional flat-top AWG, and initially design and process a flat-top Si-AWG. There are many methods to realize flat-top spectrum, such as using MMI waveguide [64, 65] or parabolic waveguide [66] between the input waveguide and FPR. This can make a double-peak mode distribution on the imaging surface of the output FPR, which superimposes the mode field of the output waveguide. As a result, the flat-top spectrum is realized. Here we used some MMI waveguides to generate the first-order mode in the waveguide. It should be mentioned that when the AWG was designed by the author for the first time, the author was not familiar with the design of MMI, so the parameters of MMI were simply evaluated by the analytical method. The final design parameters of the flat-top AWG demultiplexer with 3.2 nm channel spacing are as follows: $\Delta\lambda_{CH} = 3.2$ nm, $N_{CH} = 8$, FSR = 25.6 nm, L = 53.513μm, N = 15, $\Delta L = 25.40$, d = 1.3 μm, $d_a = 2.8$ μm, MMI length $L_{MMI} = 0.708$ μm and MMI width $W_{MMI} = 0.90$ μm. Figure 3.15a–b shows the microscopic picture of the flat-top AWG with 3.2 nm channel spacing and 8 channels. Figure 3.15c shows the normalized transmission spectrum of the AWG sample. We found that there is a modest spectral flattening at some channels, but the crosstalk performance is poor with more than –15 dB. All these problems are caused by that MMI has not been carefully optimized.

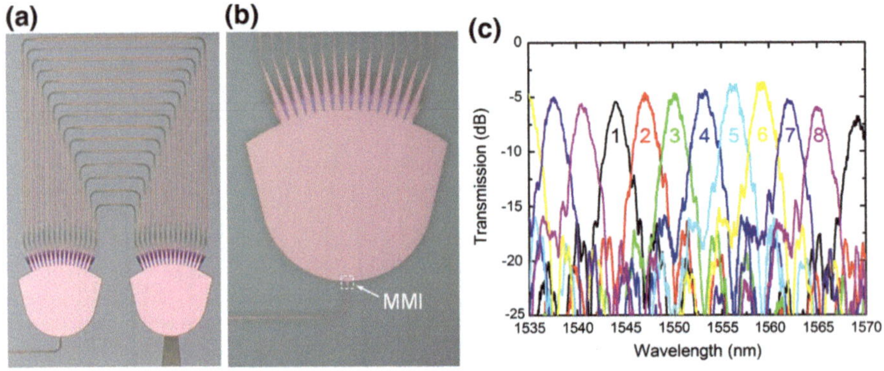

Fig. 3.15 Flat-top Si-AWG with 3.2 nm channel spacing and 8 channels. **a** Microscoped picture. **b** Local microscoped picture near the FPR. **c** Normalized transmission spectrum

3.2 PDM: Ultra-Broadband PBS Based on Cascaded Mode-Sorting Asymmetric Y-Junctions

In order to further improve the communication capacity, polarization multiplexing is a low-cost and simple multiplexing mechanism. In long-haul optical communications, polarization multiplexing can be combined with phase modulation or quadrature amplitude modulation to achieve 100 Gb/s transmission speed with single wavelength. However, due to the serious random jitter in the polarization state in the optical fiber, polarization multiplexing needs to be combined with advanced coding techniques and requires digital signal processing to decode. In the optical interconnections, the integrated optical waveguide, due to the strong birefringence, leads to excellent polarization-maintaining, which can effectively avoid the random jitter of the light polarization state, and then simplify the whole coding system. Polarization beam splitter (PBS) is one of the core devices in the polarization multiplexing mechanism. It can also realize polarization diversity mechanism with polarization rotator (PR). It has wide application prospects in polarization-insensitive photonic circuits and coherent optical communication transceivers.

For the silicon-based polarization beam splitter (Si-PBS), due to the large refractive index difference between the waveguide core layer Si and the low-refractive-index cladding layer SiO_2, it is easy to lead to strong birefringence between the TE mode and the TM mode. This phenomenon can be found in different basic optical structures, and provides various methods to realize polarization beam splitting. For example, for a PBS based on a directional coupler (DC), the coupling length or coupling efficiency is polarization dependent; for a PBS based on the multi-mode interference coupler (MMI), the beat length (L_π) is polarization dependent; for a PBS based on the MZI, the central wavelength is polarization dependent. With these polarization-dependent device parameters, various kinds of PBS devices can be realized.

3.2.1 Research Progress of Si-PBS Technology

There are many kinds of waveguide structures that can realize Si-PBS, including directional coupler (DC), multimode interference coupler (MMI), Mach-Zehnder interferometer (MZI), grating and several mode evolution structures. In recent years, surface plasmon assisted silicon waveguides have also made a series of important progress in reducing device size and increasing bandwidth. Here, we tend to briefly introduce the basic principles of various Si-PBS and review the research progress in recent years.

Because of its simple structure, PBS based on DC has received the most attention. In 2005, Kiyat et al., reported the first DC based Si-PBS [67]. They used SOI rib waveguides to realize a symmetrical DC. By designing the waveguide gap, the coupling length of TM mode is twice as long as that of TE mode. Thus, when the coupling length of TM mode is chosen, the TM mode will be completely transferred to the adjacent port, and the TE mode will return to the input waveguide after it is transferred to the adjacent waveguide. Finally, the polarization beam splitting function is realized. In 2006, Fukuda et al., implemented an ultra-small Si-PBS with only $7 \times 16\ \mu m^2$ footprint based on strip symmetric DC, and proved that the performance could be improved by cascading these PBS devices [68]. The extinction ratio of this device is better than 20 dB, and the excess loss is only 0.5 dB. In 2009, Komatsu et al., proposed a Si-PBS based on three-waveguide DC [69]. The TE mode of the middle slot waveguide is confined to the slot, and the TM mode is distributed throughout all the waveguides. So the effective refractive index of the TM mode is larger than that of the TM mode on both sides of the waveguide. When the slot waveguide is optimized to ensure the phase matching condition for the TM mode, the difference of effective refractive index of the TE modes will be larger, which will help to reduce the coupling of the TE mode and improve the extinction ratio. In 2010, Yue et al., proposed a Si-PBS based on the horizontal slot symmetric DC [70]. The length of the device is 46.7 μ m, the extinction ratio is 22 dB. The bandwidth is 18 nm around 1550 nm wavelength. In 2011, Lin et al., implemented a Si-PBS which consists of a slot waveguide and a strip waveguide [71]. The length is only 13.6 μm. For TE and TM modes, the extinction ratio are 21 dB and 17 dB respectively. In 2013, Zhang et al., realized a DC PBS which consists of a horizontal slot waveguide [72]. The test results show that the extinction ratios for TE and TM modes are better than 10 dB and 7 dB respectively in the wavelength range from 1530 nm to 1600 nm. In 2014, Kim et al., proposed a Si-PBS for polarization beam splitting between different layers for 3D optical interconnection [73]. The device, consisting of a three-waveguide DC, is only 10 μm long and has an extinction ratio better than 20 dB in the entire C-band. In 2015, Kim et al., reported a Si-PBS consisting of a three-waveguide DC [74]. For TE and TM modes, the extinction ratio/insertion loss are 22.5 dB/2.1 dB and 22.9 dB/1.8 dB, respectively. In 2015, Ying et al., put forward a Si-PBS with two waveguides with different heights and widths [75]. The lower waveguide only supports TE mode transmission, while the higher waveguide satisfies the transmission for two modes. The simulation results show that the coupling length is only 0.2 μm. For TE and TM

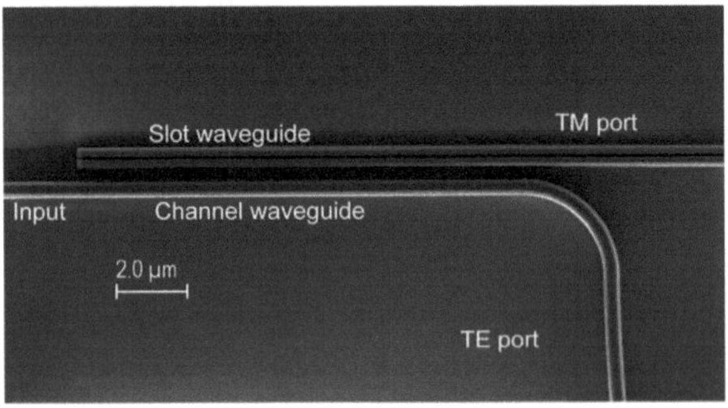

Fig. 3.16 A Si-PBS based on DC including a slot waveguide and a channel waveguide. Reprinted with permission from Ref. [71]. Copyright 2011, American Institute of Physics

modes, the extinction ratio/insertion loss are 13.5 dB/16.6 dB and 0.29 dB/0.13 dB, respectively (Fig. 3.16).

Compared with the DC structure, although MMI has larger device length, it has the advantage of large fabrication tolerance. In 2001, Rahman et al., proposed a PBS by using MMI in the III-V material platform [76]. When the length of the MMI is $L_{MMI} = n \times L_{\pi}(TE) = (n-1) \times L_{\pi}(TM)$, the input TE or TM mode output at two different ports, respectively. In 2003, Hong et al., implemented a PBC based on MMI in the silicon dioxide platform [77]. For TE and TM modes, the extinction ratio/excess loss are 11.93 dB/1 dB and 14.35 dB/0.98 dB, respectively. In 2007, Shi et al., added photonic crystals inside the MMI coupler to reduce the device length [78]. The photonic crystal can reflect one polarization state to an output port at the same propagation position as the input port without affecting the other polarizations. The device size is only $50 \times 5\ \mu m^2$. For TE and TM modes, the extinction ratio/excess loss are 27 dB/0.74 dB and 22 dB/0.54 dB, respectively. The proposal was finally demonstrated experimentally in 2010 [79]. In 2009, Jiao et al., proposed to cascaded MMI couplers to reduce the length of PBS device, which is only 1/28 of the traditional design [80]. In 2009, Yang et al., implemented Si-PBS based on two-mode interference (TMI) coupler on SOI platform for the first time [81]. The device length was only 8.8 μm. For TE and TM modes, the measured crosstalk is both about 16 dB at 1550 nm wavelength, and the extinction ratio are 18.2 dB and 13.7 dB, respectively. In 2011, Hosseini et al., further utilized TMI in the SOI platform to achieve a ultra-compact PBS with beat length of L_{π} [82]. For TE and TM modes, the extinction ratio/insertion loss are 18.2 dB/0.8 dB and 16.8 dB/1.7 dB respectively. In 2013, Huang et al., achieved a high extinction-ratio MMI-PBS [83]. The device size is $8.16 \times 1034\ \mu m^2$. For TE and TM modes, the excess loss are 1.2 dB and 2.2 dB respectively, and the extinction ratio are better than 20 dB and 15 dB in the C band, respectively. In 2014, Xu et al., proposed a PBS by using slot asymmetric MMI coupler [84]. For the TE mode, because the mode field is mainly confined in the slot region, the effect of MMI is very weak. For the TM mode, the mode field can produce self-image effect similar to that in conventional MMI coupler. The polarization splitting function can be realized by optimizing the device parameters. The simulation results show that the MMI length is only 2.3 μm. For TE and TM modes at 1550 nm wavelength, the insertion loss/extinction ratio are 1.37 dB/16.6 dB and 0.81 dB/20.9 dB, respectively. In 2015, Yin et al., realized a CMOS-compatible and fabrication-tolerant MMI-PBS [85]. The device size is $4.2 \times 132.64\ \mu m^2$. For TE and TM modes, the excess loss are 1.2 dB and 2.2 dB, respectively. The extinction ratio is better than 15 dB in 26 nm wavelength range.

The MZI structure is also often used to implement PBS. Compared with DC and MMI, MZI is more suitable for waveguide platforms with less birefringence effect and it avoids fine structure. The key to the realization of compact MZI-PBS is to introduce enough birefringence to the phase shifter. In early MZI-PBS, the stress was applied, such as depositing a layer of metal [86, 87] or amorphous silicon [88], or using some high birefringence materials, such as lithium niobate [89, 90], III-V [91], liquid crystals [92, 93], etc. For Si-PBS, the SOI waveguide itself has strong birefringence effect, so it provides a unique condition for realizing small size PBS. In 2005, Liang et al., first implemented the MZI-based Si-PBS by using SOI rib waveguides with large cross-section [94]. The length of MZI arm is 6 mm. The insertion loss is less than 6.3 dB in the wavelength range from 1520 nm to 1565 nm and the extinction ratio is better than 13 dB. In 2007, Dai et al., proposed the Si-PBS scheme by using SOI nanowire waveguides with stronger birefringence, and simulated the devices with MMI or DC as optical splitters [95]. The scheme of using MMI as a splitter was demonstrated experimentally in 2012 [96]. The total length of the device is $\sim 200\ \mu m$, and the extinction ratio is better than 10 dB in the wavelength range from 1540 nm to 1580 nm.

The grating can also realize the function of polarization beam splitting. Its advantage is that it can satisfy the function of polarization beam splitting and waveguide coupler at the same time. It solves the dependence of other PBS structures on the input and output ports which are not sensitive to polarization [97]. In 2009, Tang et al., proposed a Si-PBS by using one-dimensional gratings [98]. The coupling efficiency is $\sim 50\%$, the bandwidth is larger than 70 nm and the extinction ratio is less than –22 dB. In 2011, Wang et al., demonstrated the feasibility of one-dimensional grating PBS and introduced a bottom Bragg-grating reflector to improve the coupling efficiency [99]. The test results show that the coupling efficiency is more than 50%, the bandwidth is larger than 60 nm, and the extinction ratio is less than –20 dB. In 2013, Zaoui et al., realized a PBS by using one-dimensional gratings with bottom metal reflector [100]. The test results show that the coupling loss is less than 2.4 dB and the extinction ratio is greater than 25 dB. Theoretically, it is predicted that the coupling efficiency can be further improved by using non-uniform periodic gratings. In 2013, Streshinsky et al., first implemented a dual-wavelength PBS by using a two-dimensional grating operating at 1310 nm and 1550 nm [101]. The results show that the insertion loss/1.5 dB bandwidth are 7.1 dB/35 nm and 8.2 dB/18 nm at 1550 nm and 1310 nm wavelengths, respectively.

In order to further reduce the device size, metal can be introduced into the silicon waveguide structure to improve the birefringence effect by using the polarization dependence of the excited surface plasmon polaritons. At present, there are two main methods reported. One is the hybrid plasmonic waveguide (HPW) with an insulating layer between the metal and silicon waveguides. Specific polarization states can excite surface plasmon polaritons (SPPs), which can satisfy the requirements of low-loss, small-size and easy process. However, in practice, additional mode converters are needed between HPW and silicon waveguides. Another method is that additional metal layers or columns are added directly between silicon waveguides, which does not require additional mode converters. However, it is difficult to fabricate these converters.

In 2011, Zou et al., put forward a novel PBS by adding a silver layer in the middle of DC [102]. The input TE light can excite the SPP of the metal while the TM light travels directly through the waveguide. The simulation results show that the efficiency is as high as 92%. In 2012, Chee et al., used copper-silicon HPW to realize a PBS based on a three-waveguide DC [103]. The test results show that for TE and TM modes, the insertion loss/extinction ratio are 2.8 dB/12.3 dB and 6.0 dB/13.9 dB, respectively. In 2013, Gao et al., used silver-silicon HPW to propose a PBS based on asymmetric DC [104]. The simulation results show that the device length is 4.13 μm. For TE and TM modes, the extinction ratio are 16.4 dB and 20.9 dB, respectively. In 2013, Tan et al., proposed adding silver-metal column arrays in the DC to excite local surface plasmon (LSP) [105]. Unlike the SPP excitation, which requires the matching of the frequency and wave vector of the excitation light, LSP can be excited by the light with the appropriate frequency and polarization, independent of the wave vector of the excitation light. The simulation results show that the coupling length of the device is only 1.1 μm. For TE and TM modes, the insertion loss/extinction ratio are 0.09 dB/22.06 dB and 0.40 dB/23.06 dB, respectively. In 2014, Kim et al., proposed that CMOS-compatible copper column array can be added in the DC to stimulate LSP [106]. The length of the coupling region is only 1 μm after three-dimensional simulation, and the extinction ratio is better than 15 dB in the wavelength range of 280 nm. In 2014, Kim et al., put forward a novel PBS by using three-waveguide HPW DC for three-dimensional optical interconnections [73]. The simulation results show that the device length is 9.7 μm, and the extinction ratio in the C band can reach 30 dB. In 2014, Ma et al., proposed a PBS by using three-waveguide DC with silver-silicon HPW in the middle waveguide [107]. The simulation results show that for TE and TM modes, the insertion loss is less than 0.5 dB and extinction ratio is greater than 14 dB in C the band. In 2015, Xu et al., proposed a PBS by using three-waveguide DC with silver-silicon HPW on the two sides [108]. The TE and TM mode input from the middle silicon waveguide will only be coupled to the horizontal and vertical HPW, respectively. The simulation results show that for TE and TM modes, the coupling length/best extinction ration/loss are 6.5 μm/18.9 dB/0.44 dB and 4.5 μm/15.2 dB/0.89 dB, respectively (Fig. 3.17).

It is worth mentioning that the Dai's group from Zhejiang University and his collaborators have made a series of important achievements in the study of Si-PBS. In addition to the work based on MZI [95, 96] and grating structure [98] introduced previously, it also includes some other devices. In 2011, an ultra-small Si-PBS was proposed by using asymmetric DC consisting of a strip waveguide and a slot waveguide, in which a slot waveguide can use a S-bend waveguide to achieve mode conversion [109]. The simulation results show that the device length is only 6.9 μm, while the coupling length is only 1.3 μm. In 2011, an ultra-small Si-PBS was pro-

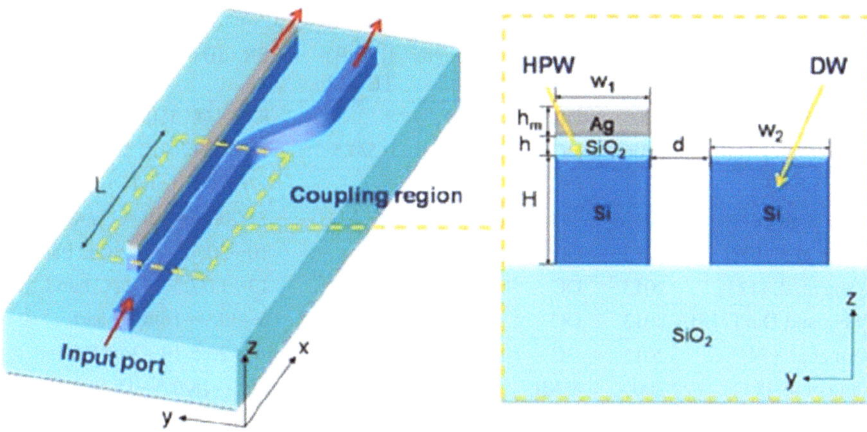

Fig. 3.17 A Si-PBS based on Silver-Si HPW. Reprinted from Ref. [104], Copyright 2013, with permission from Elsevier

posed by using a bend asymmetric DC. The device size is less than 10 μm [110]. In 2013, such structures were experimentally demonstrated to have an extinction ratios of better than 10 dB over a wide wavelength range [111]. In 2012, a Si-PBS was proposed by using a three-waveguide asymmetric DC structure with a wider middle waveguide [112]. The input TM_0 mode can be converted to the TM_1 mode in the middle wide waveguide, and then to the TM_0 mode in the output waveguide. The input TE_0 mode can not be coupled with the adjacent waveguide. The length of the device is about 25 μm, and the extinction ratio is better than 15 dB in 50 nm wavelength range. In 2013, such a structure was experimentally demonstrated, and the test results showed that the extinction ratios for the TE and TM modes are 11 dB and 13 dB respectively in 42 nm wavelength range [113]. In 2012, a PBS was proposed by using a three-waveguide DC structure with silver-silicon HPW in the middle [114]. The simulation results show that the device size is only $2\times5.1\ \mu m^2$, the insertion loss in the C band is less than 1.5 dB, and the extinction ratio is more than 15 dB. In 2013, a PBS was proposed by using an asymmetric DC structure based on silver-silicon HPW [115]. The simulation results show that the device size is only $1.9\times3.7\ \mu m^2$. The insertion loss are 0.025 dB and 0.66 dB for TE and TM modes, respectively. The extinction ratio is better than 12 dB over a wavelength range of 120 nm. In 2014, a PBS was proposed by using silver-silicon HPW MMI structure [116]. The MMI area is partly covered by metal, which only produces a self-image effect on the TE mode, but not on the TM mode. The simulation results show that the device size is only $1.8\times2.5\ \mu m^2$. The loss is less than 1 dB and the extinction ratio is better than 10 dB in the wavelength range of 80 nm.

The basic structure of the PBS mentioned above belongs to the mode interference structure. Generally speaking, it is more sensitive to the size of the device and the bandwidth is limited. In order to overcome this problem, PBS can be realized by using mode-evolution structure or wavelength-insensitive structure. In 2005, Watts

Table 3.4 Experimental summary of Si-PBS (2005–2015.5)

Group	Time	Structure	Length	IL (dB) TE/TM	ER (dB) TE/TM
Kiyat et al. [67]	2005	DC	120	*	8/18.1@1550 nm
Fukuda et al. [68]	2006	DC	16	0.5/0.5	11~16/11~12.5@C band
Lin et al. [71]	2011	DC	15.8	*	17/21@C band
Chee et al. [103]	2012	DC	9.5	4/7	7~20/14~30@C band
Zhang et al. [72]	2013	DC	20	*	10~20/7.5~17.5@C band
Wang et al. [111]	2013	DC	10.1	*	13~17/7.5~17@C band
Wang and Dai [113]	2013	DC	25.5	*	5~30/9~16@C band
Kim et al. [74]	2015	DC	12.5	2.1/1.8	18~26.5/19~24@C band
Yang et al. [81]	2009	MMI	8.8	0.5/2	15.5~18/9.5~15.5@C band
Hosseini et al. [82]	2011	MMI	0.94	0.8/1.7	14~22.5/15~18.5@C band
Huang et al. [83]	2013	MMI	1034	*	>20/10~25@C band
Yin et al. [85]	2015	MMI	132.64	1.2/2.2	11~22.7/12.5~30.6@C band
Liang and Tsang [94]	2005	MZI	6500	<6.3/<6.3	12.5~19/14~19@C band
Dai et al. [96]	2012	MZI	200	*	8.5~17/14~23@C band

*denotes no performance characterization from the reference in the table

et al., proposed a PBS based on mode evolution [117]. The 3D-FDTD simulation results show that the conversion efficiency increases with the device length, and the conversion efficiency exhibits the ultra-broadband properties of several hundred nanometers. In 2014, Su et al., implemented a four-port Si-PBS based on the mode evolution structure [118]. The test results show that the crosstalk of the device for 150 nm bandwidth of 150 nm is less than –10 dB. In 2014, Uematsu et al., proposed using wavelength-insensitive couplers to achieve ultra-broadband Si-PBS [119]. The simulation results show that the mode conversion efficiency is higher than 90%. The extinction ratio are better than 17 dB and 13 dB for TE and TM modes respectively in the wavelength range of 200 nm.

Finally, we summerize the experimental results of Si-PBS based on DC, MMI and MZI structures, as shown in Table 3.4.

3.2.2 *Ultra-Broadband Silicon-on-Insulator Polarization Beam Splitter Based on Cascaded Mode-Sorting Asymmetric Y-Junctions*

Y-junction is a very important basic device in integrated optics. The Y-junction with symmetric arm are often used as optical splitter, and have been extensively studied for many years. However, when the width of the Y-junction arm is different, that is, for the asymmetric Y-junction, the mode-sorting effect will occur. In general, when the length of the asymmetric Y-junction can satisfy the adiabatic mode conversion,

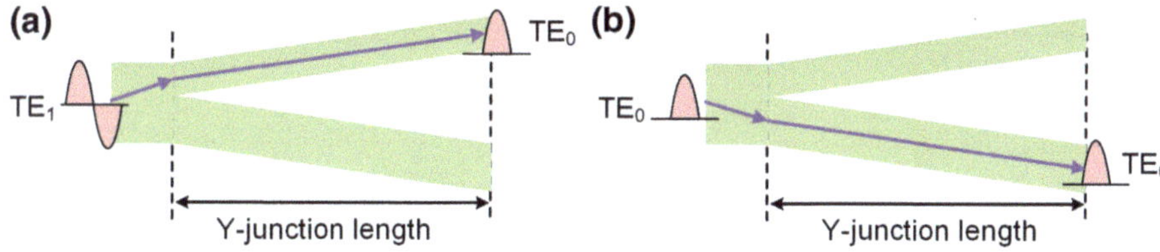

Fig. 3.18 a–b Schematic of the function of an asymmetrical Y-junction for TE_1 and TE_0 inputs, respectively

the light from the root waveguide will output in a certain mode in a certain arm waveguide, and the effective refractive index of this mode must be closest to that of the input mode. The same principle is also applied to the case of light coming from the arm waveguide.

Figure 3.18a–b show the functional diagram in an asymmetric Y junction for TE_1 and TE_0 inputs, respectively. When the TE_0 mode inputs from the root waveguide, it must output from the wide arm waveguide because it must have the closest effective refractive index with the TE_0 mode in the wide arm waveguide. When the TE_1 mode inputs from the root waveguide, as long as the TE_0 mode of the narrow arm waveguide has the closest effective refractive index to the TE_1 mode of the root waveguide, it will output from the narrow arm waveguide as the TE_0 mode. Of course, if the width of the narrow arm waveguide is too narrow, the effective refractive index of the TE_1 mode in the root waveguide and wide arm waveguide are the closest, the light will output in the wide arm waveguide as the TE_1 mode.

Based on the mode-sorting effect, asymmetric Y-junctions can not only provide new ideas for the design of integrated optical devices, but also have the advantages of large bandwidth and high fabrication tolerance. This is mainly because the mode conversion is based on mode evolution rather than mode interference. The researchers at the Australian National University first studied the mode-sorting effect in a variety of asymmetric Y-junction waveguides [120], and proposed some principles for the design of asymmetric Y-junction [121]. After that, there are also a lot of reports, which make use of asymmetric Y-junctions to realize mode division multiplexing [122–125]. However, the mode-sorting effect in the above asymmetric Y-junction is mainly used for TE modes. By carefully selecting the width of the arm waveguide and using cascaded structure, a more complex mode sorting between TE and TM modes can be achieved. One of the typical applications is PBS.

Figure 3.19 is the schematic of our proposed ultra-broadband Si-PBS based on cascaded asymmetrical Y-junctions. The whole device is based on a SOI strip waveguide with a 220 nm top silicon layer and SiO_2 upper cladding. The whole device consists of three asymmetric Y-junctions. The N-th Y-junction is composed of a root waveguide, a narrow arm waveguide and a wide arm waveguide, with a width of W_{yNs}, W_{yNn}, and W_{yNw}, respectively. The length is L_{yN}. If the widths of the two adjacent root waveguides are different, the root waveguides can be connected by a tapered mode converter.

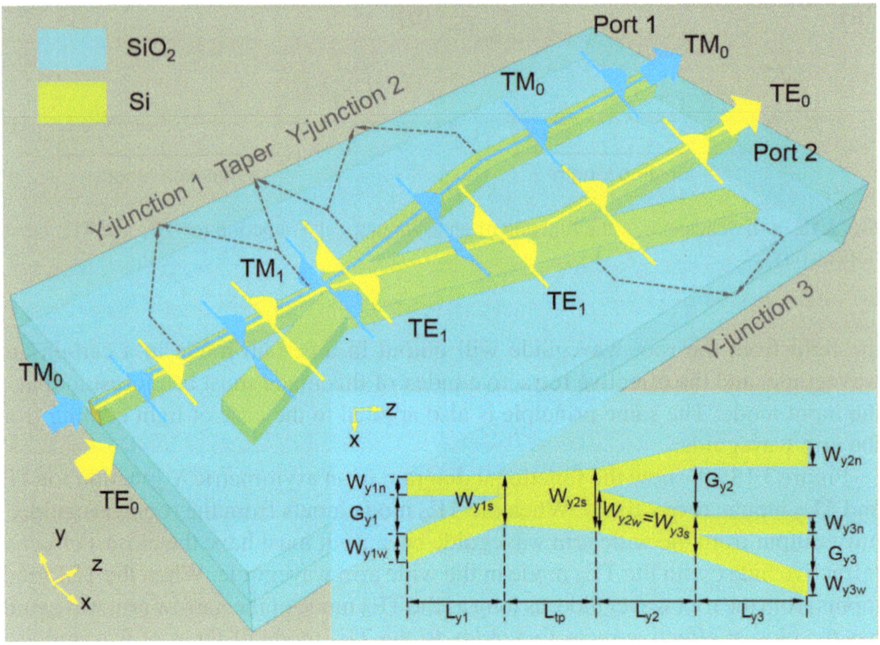

Fig. 3.19 Schematic of our proposed ultra-broadband Si-PBS based on cascaded asymmetrical Y-junctions. [2014] IEEE. Reprinted, with permission, from Ref. [126]

The operation principle of the overall device is described as follows

- For the TM_0 input, it enters the device from the narrow arm waveguide of the first Y-junction, and gradually evolves into the TM_1 mode of the root waveguide. Then, the TM_1 mode will evolve into the TM_0 mode of narrow waveguide of the second Y-junction, and finally output in port 1. Here, we can assume that the TM_0 mode inputs in the wide arm of the first Y-junction, and that the TM_0 mode must evolve into a TM_0 mode in the root waveguide, because the effective refractive index of the two modes is the closest. This is an ineffective operation for polarization splitting, so we choose the narrow arm as the input waveguide.
- For the TE_0 input, it enters the device from the narrow arm waveguide of the first Y-junction, and gradually evolves into the TE_1 mode of the root waveguide. Then, the TE_1 mode is evolved into the TE_1 mode of the wide arm of the second Y-junction. Finally, the TE_1 mode is evolved into the TE_0 mode of the narrow arm of the third Y-junction and outputs at port 2. Based on the same consideration, it is still not possible to select the wide arm in the first Y-junction as the input waveguide.

In order to realize mode-sorting function in each Y-junction above, we first used the two-dimensional simulation software FIMMWAVE to calculate the effective refractive index of the five guided modes (TE_0, TE_1, TM_0, TM_1 and TE_2) supported in the

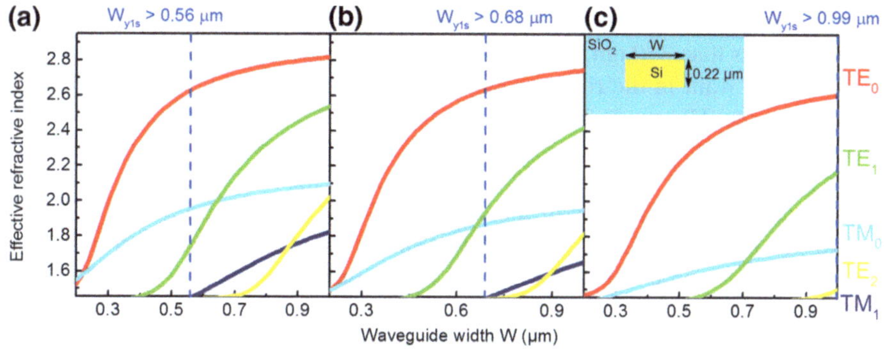

Fig. 3.20 **a–c** Calculated effective refractive indices of the first five modes as a function of the waveguide width in a strip waveguide at wavelengths of 1450 nm, 1550 nm, and 1750 nm, respectively. [2014] IEEE. Reprinted, with permission, from Ref. [126]

strip waveguide. Figure 3.20a–c shows the relationship between the effective refractive index of waveguide mode and the waveguide width W when the wavelengths are 1450 nm, 1550 nm and 1750 nm, respectively. Considering that the wavelength range in this device is large, we use the wavelength-dependent models of Si and SiO_2 given by FIMMWAVE, so as to ensure the accuracy of calculation. According to the working principle of the device, the root waveguide of the first Y-junction needs to support at least four guided modes (TE_0, TE_1, TM_0 and TM_1). From Fig. 3.20, we can see that the waveguide width W_{y1s} of the first Y-junction need to be larger than 0.56 μm, 0.68 μm and 0.99 μm for wavelengths of 1450 nm, 1550 nm and 1750 nm, respectively. Therefore, in order to ensure the device working well in the wavelength range from 1450 nm to 1750 nm, we choose $W_{y1s} = 1\,\mu m$.

For the design and optimization of each Y-junction, we used the 3D simulation software FIMMPROP based on the eigenmode expansion method to calculate the mode conversion efficiency. This part of the simulation is performed at the wavelength of 1550 nm.

1. Design and optimization of the first Y-junction.
   Figure 3.21a–b shows the mode conversion efficiency between the TE_0/TM_0 mode from the narrow arm to the TE_1/TM_1 mode from the root waveguide with different narrow arm widths W_{y1n}. The gap between these two arm waveguides is $G_{y1} = 2\,\mu m$ at the input. For these two polarizations, the mode efficiency increases with the increase of the Y-junction length L_{y1}. However, because there is still a certain difference between the effective index of the input and output mode, there will be some scattering or reflection at the corner of the Y-junction and the arm waveguide, causing the mode conversion efficiency to be slightly smaller than 100%. It can be found that as long as the Y-junction length L_{y1} is large enough, when the width of the narrow arm is between 330 nm and 490 nm, the exact mode-sorting can always be obtained, and the conversion efficiency is high enough. This phenomenon also indicates that the device has high

fabrication tolerance potentially. When the waveguide width of the narrow arm W_{y1n} is less than 330 nm, the input TE_0 will be converted into the TE_2 mode that we do not want in the root waveguide. This is because the effective refractive index of the TE_1 mode in the wide arm waveguide is closer to the TE_1 mode in the root waveguide than the TE_0 mode in the narrow arm waveguide. Here, we select $W_{y1n} = 0.41\mu m$ and $L_{y1} = 200\ \mu m$ for the first Y-junction, so that the high conversion efficiency of more than 96% can be obtained for both of these two input polarizations. Figure 3.21c–d shows the mode propagation in the first Y-junction for both of the two input polarizations.

2. Design and optimization of the second Y-junction.
   In order to avoid using a mode converter between first Y-junction and second Y-junction, we made the root waveguide width of the second Y-junction $W_{y2s} = W_{y1s} = 1\mu m$. The function of the second Y-junction is to convert the TM_1 input from the root waveguide to the TM_0 mode of the narrow arm waveguide, while maintaining the TE_1 mode into the wide arm waveguide. Figure 3.22a–b show the relationship between the conversion efficiency from the TE_1/TM_1 mode in the root to the TE_1/TM_0 mode in the wide/narrow arm respectively when the width W_{y2n} of the narrow arm waveguide is from 250 nm to 310 nm. Here, the gap between the two arm at the output end is set to $G_{y2} = 3\mu m$ to reduce the coupling between the output waveguides and to improve the polarization extinction ratio (PER) of the device. It can be seen from the figure that when the Y-junction length is relatively small, the efficiency of mode conversion does not always increase with the increase of Y-junction length L_{y2}. This may be because some input power is coupled to multiple modes in the arm waveguide, resulting in some mode interference or reflection. Here, we select $W_{y2n} = 0.27\mu m$ and $L_{y2} = 200\mu m$ for the second Y-junction to guarantee high mode conversion efficiency. At this time, the width of the corresponding wide arm is $W_{y2w} = 0.73\mu m$, which is large enough to support 3 guided modes, which will contribute to the design of the third Y-junction. At the same time, there is no need to add an assisted tapered mode converter between the second Y-junction and third Y-junction. Figure 3.22c–d show the mode propagation of the TE_1/TM_1 input mode from the root waveguide in the optimized second Y-junction, respectively.
3. Design and optimization of the third Y-junction.
   Figure 3.23a shows the relationship between the mode conversion efficiency from the TE_1 mode in the root waveguide to the TE_0/TM_0 modes in the narrow arm, and the Y-junction length L_{y3} with different narrow arm widths W_{y3n}. Here, the gap between the two arms at the output is $G_{y3} = 1\mu m$. Because the TM_0 mode has been separated from the second Y-junction, we only need to consider the TE_0 mode in the Y-junction. Here, we choose $W_{y3n} = 0.35\mu m$ and $L_{y3} = 200\mu m$ for third Y-junction to guarantee the mode conversion efficiency of about 99%. Figure 3.23b shows the propagation of the TE_1 mode from the root waveguide in the optimized third Y-junction.

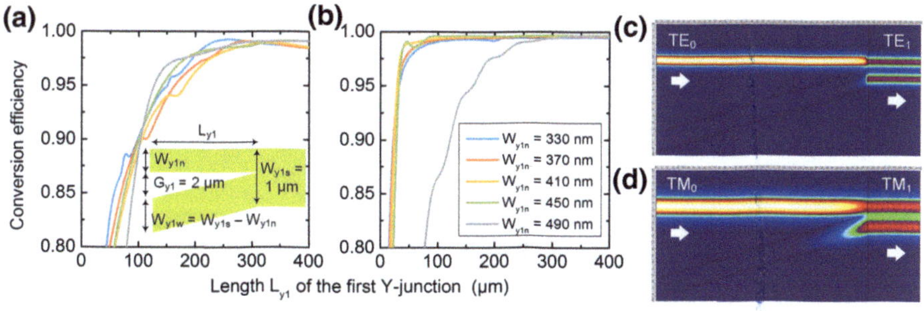

Fig. 3.21 Optimization of the first Y-junction. **a**–**b** Mode conversion efficiency from TE_0/TM_0 in the narrow waveguide to TE_1/TM_1 in the stem. **c**–**d** Mode propagation in the first Y-junction. [2014] IEEE. Reprinted, with permission, from Ref. [126]

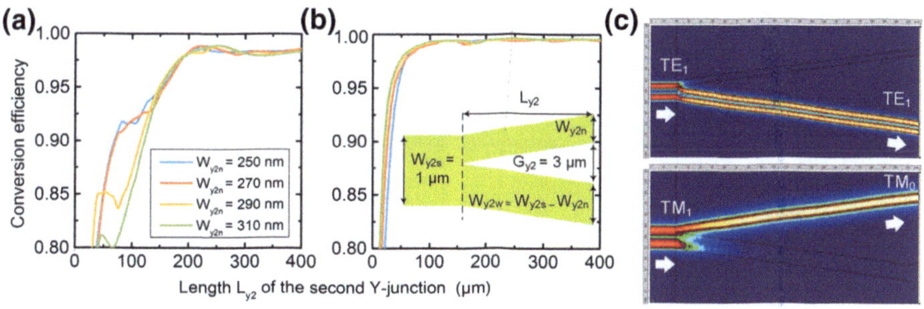

Fig. 3.22 Optimization of the second Y-junction. **a**–**b** Mode conversion efficiency from TE_1/TM_1 in the stem to TE_1/TM_0 in the wide/narrow arm. **c**–**d** Mode propagation in the second Y-junction. [2014] IEEE. Reprinted, with permission, from Ref. [126]

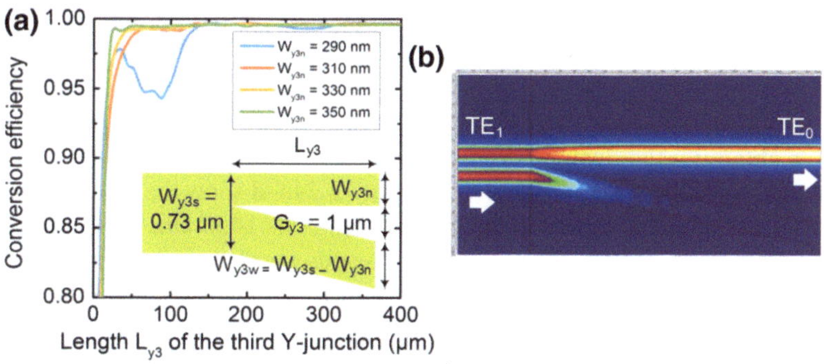

Fig. 3.23 Optimization of the third Y-junction. **a** Mode conversion efficiency from TE_1 in the stem to TE_0 in the narrow arm. **b** Mode propagation in the third Y-junction. [2014] IEEE. Reprinted, with permission, from Ref. [126]

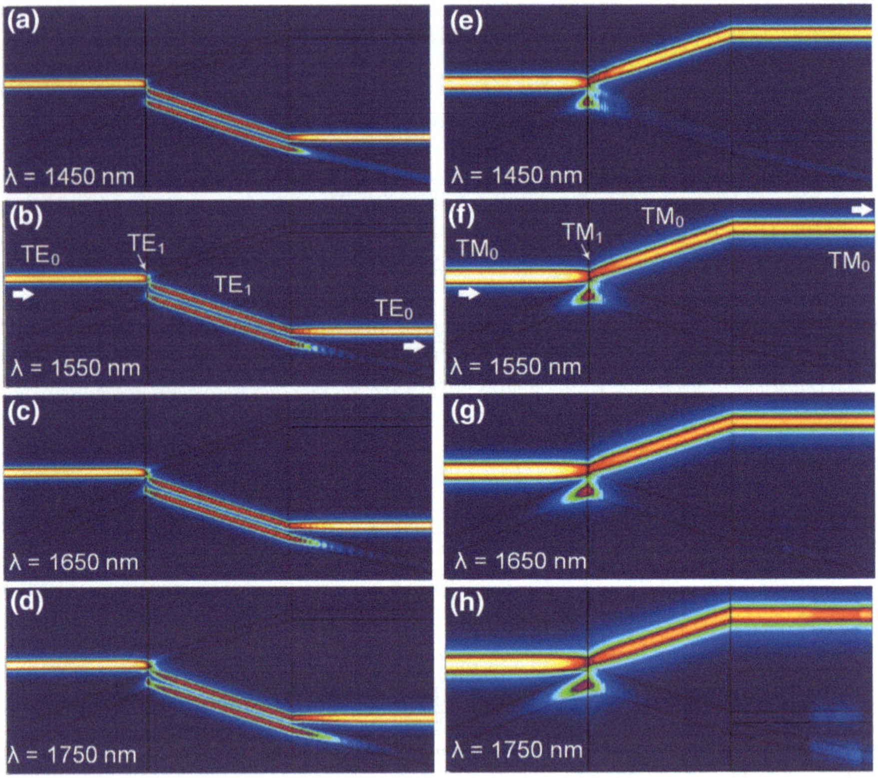

Fig. 3.24 Mode propagation in the PBS for the incoming TE_0 mode at wavelength of **a** 1450 nm, **b** 1550 nm, **c** 1650 nm and **d** 1750 nm. Mode propagation in the PBS for incoming TM_0 mode at wavelength of **e** 1450 nm, **f** 1550 nm, **g** 1650 nm and **h** 1750 nm. [2014] IEEE. Reprinted, with permission, from Ref. [126]

After obtaining the design parameters of each Y-junction, we further carry out the simulation of the whole PBS device. Figure 3.24a–d show the propagation of the TE_0 mode in the whole device when the wavelength is 1450 nm, 1550 nm, 1650 nm and 1750 nm, respectively. The simulation results show that the TE_0 mode can be effectively separated from the correct ports at different wavelengths. Figure 3.24e–h show the propagation of the TM_0 mode in the whole device when the wavelength is 1450 nm, 1550 nm, 1650 nm and 1750 nm, respectively. We can see that at long wavelengths, part of the power is coupled to other ports. This is because at a longer wavelength, the optical mode limitation in the waveguide is weakened, resulting in a stronger coupling between port 1 and port 2. To reduce such coupling, a S-bend waveguide near the port 1 can be used to increase the port gap.

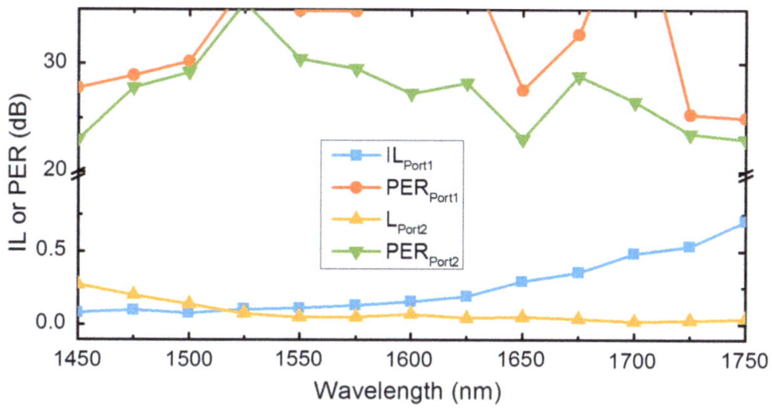

Fig. 3.25 IL and PER as a function of the wavelength in different output ports. [2014] IEEE. Reprinted, with permission, from Ref. [126]

To characterize the device performance, we define insertion loss (IL) and polarization extinction ratio (PER) of each output port as follows:

$$\mathrm{IL_{Port1}} = -10\log_{10}(\mathrm{P}^{\mathrm{TM_0}}_{\mathrm{Port1,TM_0}}) \tag{3.7}$$

$$\mathrm{IL_{Port2}} = -10\log_{10}(\mathrm{P}^{\mathrm{TE_0}}_{\mathrm{Port2,TE_0}}) \tag{3.8}$$

$$\mathrm{PER_{Port1}} = 10\log_{10}(\mathrm{P}^{\mathrm{TM_0}}_{\mathrm{Port1,TM_0}}/\mathrm{P}^{\mathrm{TM_0}}_{\mathrm{Port1,TE_0}}) \tag{3.9}$$

$$\mathrm{PER_{Port2}} = 10\log_{10}(\mathrm{P}^{\mathrm{TE_0}}_{\mathrm{Port2,TE_0}}/\mathrm{P}^{\mathrm{TE_0}}_{\mathrm{Port2,TM_0}}) \tag{3.10}$$

Here, $\mathrm{P}^{i}_{j,k}$ is the normalized power when the input mode is i and output mode is k at output port j. Some people define IL and PER according to polarization state because the input $\mathrm{TM_0}$ and $\mathrm{TE_0}$ modes correspond to the output at port 1 and port 2, respectively.

Figure 3.25 shows the wavelength dependence of the device performance. Because this device mainly uses mode evolution instead of mode interference, it shows very good IL and PER performances. In the wavelength range from 1450 nm to 1750 nm, these two ports can achieve less than 0.7 dB IL and better than 22 dB PER, covering the S, C, L and U communication bands.

We further analyzed the fabrication tolerances of this device. Figure 3.26a–b show the effects of the two most common process errors (width error $\Delta W_{\mathrm{device}}$ and the height error $\Delta H_{\mathrm{device}}$) on the performance of the device, respectively. When $\Delta W_{\mathrm{device}}$ changes between –20 and 30 nm, IL is still less than 1 dB, and PER is higher than 20 dB. Moreover, device performance is also not sensitive to height error $\Delta H_{\mathrm{device}}$. Even if $\Delta H_{\mathrm{device}}$ changes by ±30 nm, IL is still less than 0.4 dB, and PER is better than 24 dB.

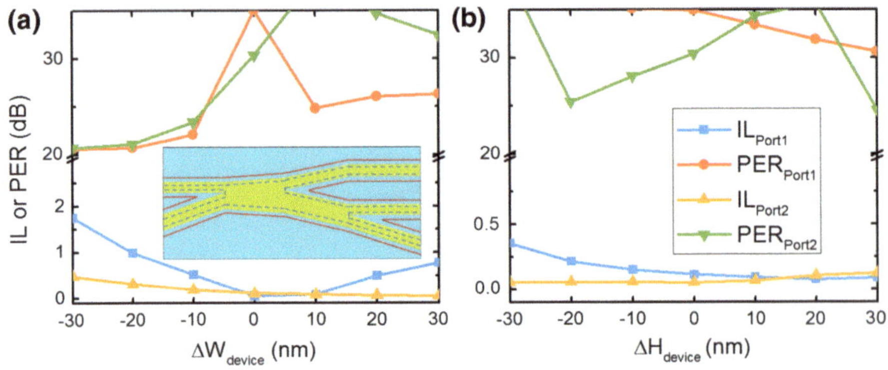

Fig. 3.26 Fabrication tolerance for the overall PBS at 1550 nm wavelength with respect to the deviations on **a** width ΔW_{device} and **b** height ΔH_{device}, respectively. The inset of (**a**) shows a PBS layout affected by the width deviation ΔW_{device}. [2014] IEEE. Reprinted, with permission, from Ref. [126]

Although the simulation results show that the PBS device has excellent performance and high tolerance, the following problems need to be considered in practical fabrication.

- Firstly, in addition to the mode conversion loss shown in the simulation, the scattering loss caused by the surface roughness of the waveguide has great effect on the overall loss of the device. This loss is mainly reduced by improving lithography and etching processes. In the previous work of our group, we made a systematic study on how to achieve low-loss silicon optical waveguides [127].
- Secondly, the sharp corner between the Y-junction arms needs to be improved. The sharp corner of the Y-junction will cause some reflection loss in practice, while the flat corner or rounded corner may destroy the mode-sorting function we need. Fortunately, the mode-division-multiplexing mechanism has recently been reported by using mode-sorting function in the asymmetrical Y-junction, and the crosstalk is as high as –30 dB [122, 123]. And some people have proved some ways to improve the Y-junction loss [128]. Therefore, we think that this PBS can be fabricated in practice.
- Finally, one disadvantage of this device is that the device length is longer than that based on DC or MMI in the same material system. Although our simulation results show that the device performance improves with the increase of device length, the waveguide propagation loss increases with the increase of length. Especially in this device, the widths of some arm waveguides are only a slightly larger than 200 nm, which makes most of the optical mode field not limited in the waveguide core and the scattering loss is increased. Considering this point, the actual device length with the best performance should be far less than 600 μm.

3.3 MDM: Broadband and Fabrication-Tolerant MDM Based on Counter-Tapered Couplers

Space division multiplexing is also a multiplexing technology that can effectively improve the communication capacity. At present, it is mainly divided into the following two categories [129].

- In the field of optical fiber communication [130], multi-core optical fiber (MCF) can be adopted [131]. Its advantage is that different channels have been physically separated without any multiplexer or demultiplexer. The disadvantage is that relatively large device area, especially when the number of channels reaches a certain degree, will bring great difficulty on the link design. For example, some optical crossing waveguides will be introduced, resulting in additional losses. Moreover, limited to the physical area, the signal conversion between different waveguides is also difficult. On the other hand, few mode optical fibers (FMF) can be adopted, which is also the current research hotspot [132–138]. Its advantage lies in the use of various intrinsic modes in optical fiber for transmission. However, it requires a more complex spatial light testing platform for mode multiplexing and demultiplexing, so it is more difficult to expand to more modes.
- In the field of on-chip optical interconnection, multi-mode optical waveguides are adopted [129]. Multiplexing and demultiplexing between different eigenmodes can be performed by using special integrated optical devices, so it is relatively simple. The focus of our research is also on the mode division multiplexing technology of silicon-based optical interconnection.

3.3.1 Research Progress of Si-MDM Technology

The design and implementation of mode division multiplexing on silicon substrate (Si-MDM) can be based on various basic optical structures, including asymmetric directional coupler (ADC), multi-mode interference coupler (MMI), asymmetric Y-junction and etc. In this part, we mainly review the recent research progress of Si-MDM.

The Si-MDM based on ADC is simple in structure, clear in design, small in size, and can be easily combined with WDM devices and PDM devices to achieve more complex hybrid multiplexing mechanism. In 2013, Dai et al., first demonstrated a Si-MDM based on ADC [139]. By designing the width of two waveguides in ADC, the TM_0 mode can be coupled to other higher order TM modes in the multimode transmission waveguide. The test results show that the whole link has less than 1 dB excess loss and less than –23 dB crosstalk in the wavelength range of 20 nm. In 2014, Wang et al. used ADC to realize an 8-channel PDM-MDM hybrid multiplexing [140]. PDM uses a PBS based on three-waveguide DC. The test results show that the crosstalk is at –10 dB~–15 dB in the wavelength range of 30 nm. In 2014, Wang et al., used grating couplers as polarization filters to further improve the crosstalk performance of an 8-channel PDM-MDM link [141]. The test results show that the crosstalk is about –20 dB in the wavelength range of 100 nm. In 2014, Wang et al., combined ADC-based 4-mode Si-MDM with four 16-channel AWGs to implement a 64-channel WDM-MDM hybrid multiplexing mechanism [142]. The test results show that the loss and crosstalk are around 7 dB and –10 dB, respectively. In 2015, Dai et al., combined ADC-based 4-mode Si-MDM with two bidirectional AWGs to implement a 64-channel WDM-MDM hybrid multiplexing mechanism [143]. With the use of bidirectional AWG, the area of the device can be reduced by nearly half. The test results show that the excess loss is 3.5 dB $\sim$ 5 dB, the wavelength channel crosstalk is –14 dB, and the mode channel crosstalk is –20 dB. In 2014, Luo et al., implemented WDM-MDM hybrid multiplexing mechanism by combining ADC with thermally controlled micro-ring filters, and carried out high-speed system experiments [144]. In 2013, Ding et al., implemented fabrication-tolerant Si-MDM with TE_0 and TE_1 modes by using tapered waveguide ADC [145]. The test results show that the insertion loss is only 0.3 dB, and the crosstalk is less than –16 dB in 100 nm wavelength range. In 2013, Qiu et al., proposed a 4-mode Si-MDM scheme by using grating-assisted contra-ADC [146]. The mode conversion efficiency between grating-assisted contra-ADC waveguides is not periodic as that of conventional ADC waveguides, but increases with the length of coupling region, which can reduce the difficulty of fabrication. Moreover, grating-assisted ADC has the characteristics of WDM, so it can realize the hybrid multiplexing mechanism of MDM and WDM. In 2015, the Si-MDM scheme of grating-assisted inverse ADC was demonstrated by Chen et al. [147].

MDM based on MMI has received considerable attention in recent years because of its large fabrication tolerance. In 2012, Uematsu et al., proposed a two-mode MDM scheme by using cascaded MMI couplers and phase shifters [148]. The simulation results show that the device works well and has large process tolerance in the C band. In 2014, Li et al., used a MMI coupler and Y-junction splitter to propose a two-mode MDM [149]. The simulation results show that the crosstalk is less than –22 dB in the wavelength range from 1500 nm to 1600 nm. In 2014, Ye et al., realized the conversion from PDM to a two-mode MDM by using two-dimensional grating PBS and cascaded MMI couplers [150]. The test results showed that for TE and TM inputs, the extinction ratio are 16 dB and 20 dB at 1550 nm wavelength, respectively.

In addition to mode interference structures such as ADC and MMI, MDM can also be implemented by using the mode-sorting of asymmetric Y-junctions. Mode conversion in asymmetric Y-junctions is based on mode evolution, so it has the characteristics of broadband bandwidth and high process tolerance. In 2013, Chen et al., proposed a MDM scheme by using the mode-sorting effect of asymmetric Y-junctions [124]. The simulation results show that the crosstalk is less than –21.8 dB in the wavelength range of 140 nm. In 2013, Driscoll et al., implemented a two-mode Si-MDM based on asymmetric Y-junctions [122]. The test results show that the crosstalk in the C band is less than –9 dB and the insertion loss is less than 1.5 dB. In 2014, Driscoll et al., verified the above Si-MDM by high-speed WDM system [123]. The test results show that 60 Gb/s transmission rate can be achieved in a 3-channel

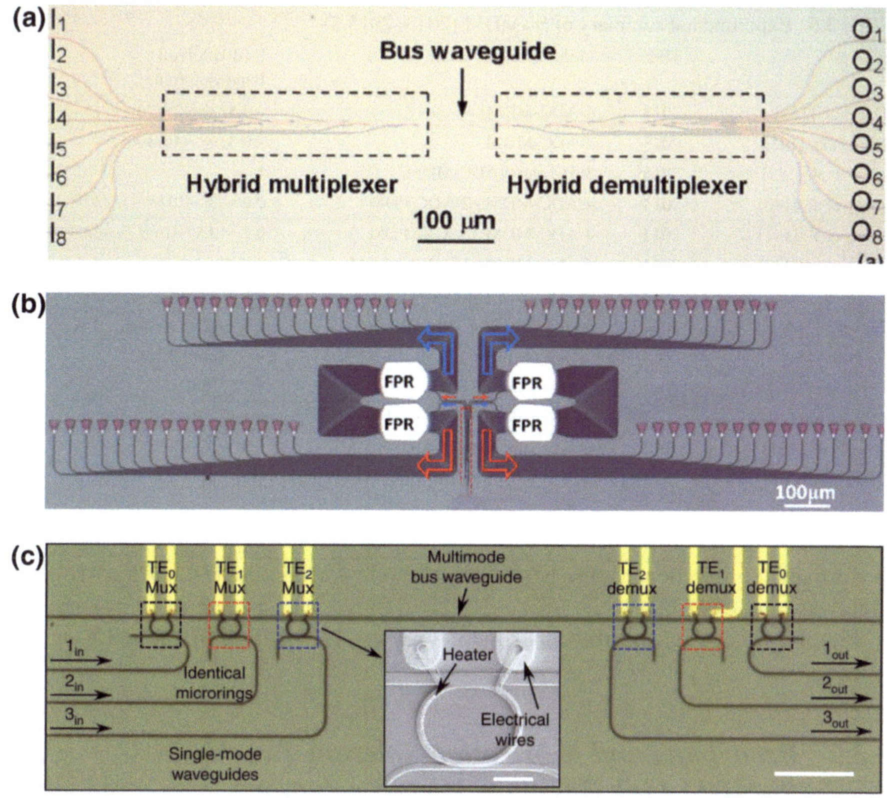

Fig. 3.27 MDM mechanism based on ADC. **a** ADC-PDM-MDM. Reproduced from Ref. [140] by permission of John Wiley & Sons Ltd. **b** ADC-MDM+AWG-WDM. Reproduced from Ref. [143] by permission of John Wiley & Sons Ltd. **c** ADC-MDM+Ring-WDM. Reprinted by permission from Macmillan Publishers Ltd: Ref. [144], copyright 2014

WDM and 2-channel MDM hybrid multiplexing link, and the power compensation of each channel is less than 0.7 dB (Fig. 3.27).

In addition to the above MDM mechanism, there are other work worth mentioning. In 2013, Xing et al., implemented broadband and fabrication-tolerant two-mode Si-MDM by using adiabatic couplers [151]. The crosstalk is less than –36 dB in the wavelength range from 1500 nm to 1600 nm, and the insertion loss is about 0.3 dB. In 2014, Dorin et al., studied the application of micro-ring resonator as mode up/down filter in the MDM [152]. This methord can effectively reduce the MDM link area and maintain good crosstalk performance. In 2014, Chen et al., used the asymmetric Y-junctions to propose an MDM multiplexer/demultiplexer [125]. The simulation results show that the device can achieve less than 0.13 dB excess loss, more than 28.4 dB extinction ratio and less than –25.5 dB crosstalk in the wavelength range of 210 nm. In 2014, Dai et al., investigate a new method to improve the

Table 3.5 Experimental summary of Si-MDM (2013–2015.5)

Group	Time	Multiplexing structure	IL(dB)/XT(dB)/ Bandwidth(nm)
Dai et al. [139]	2013	4-ADC-MDM	$<0.5/<-23/20$
Ding et al. [145]	2013	2-ADC-MDM	$<0.3/<-16/100$
Chen et al. [147]	2015	3(5)-GratingADC-MDM	*
Wang et al. [140]	2014	4-ADC-MDM×2-ADC-PDM	$0.2\sim 2/-10\sim -15/30$
Wang et al. [141]	2014	4-ADC-MDM×2-ADC-PDM	$0.2\sim 3.5/-16.6\sim -38.3/100$
Wang et al. [142]	2014	4-ADC-MDM×16-AWG-WDM	$7/-10/*$
Dai et al. [143]	2015	4-ADC-MDM×16-AWG-WDM	$3.5\sim 5/-14\sim -20/*$
Luo et al. [144]	2014	3-ADC-MDM×3-Ring-WDM	$1.5/-12\sim -22/*$
Ye et al. [150]	2014	2-Grating-PDM->2-MMI-MDM	$*/-16\sim -20/*$
Driscoll et al. [122]	2013	2-Y_junction-MDM	$<1.5/-30/35$
Driscoll et al. [123]	2013	2-Y_junction-MDM×3-WDM	$<1.5/-30/35$
Xing et al. [151]	2013	2-Adiabatic_coupler-MDM	$0.3/-36/100$

*denotes no performance characterization from the reference in the table

performance of the bend in the MDM multi-mode waveguides by using narrow and high waveguides [153].

Finally, we summarize the experimental results of Si-MDM in Table 3.5.

3.3.2 Broadband and Fabrication-Tolerant Two-Mode Si-MDM Link Based on Mode-Evolution Counter-Tapered Couplers

So far, Si-MDM has been successfully realized in different optical structures. But carefully analyzing the reported Si-MDM, we can find that for Si-MDM based on general ADC, strict phase matching conditions are required, so the device performance is more sensitive to size error, resulting in a bandwidth of only a few nanometers [139–142, 144]. For Si-MDM based on tapered ADC, there is only a two-mode MDM reported [145]. For Si-MDM based on grating-assisted ADC, a high-quality grating structure is needed [145, 147]. For Si-MDM based on MMI, its structure is relatively complex and requires a combination of a lot of basic optical structures, so it is difficult to extend to MDM [148–150] over two modes. For Si-MDM based on asymmetric Y-junctions, the sharp angle between the arm and the root waveguides must be carefully optimized, and the relationship between the mode conversion loss and the waveguide scattering loss need to be considered properly [122–124]. For MDM based on adiabatic mode coupler, it is also difficult to extend to MDM with two or more modes [151].

In order to satisfy the requirements of large bandwidth, high fabrication tolerance and extensibility in a high performance MDM, it is proposed that the counter-tapered coupler can be used as mode-selection device [154, 155], and has been proved in the small mode fiber (FMF) [134, 138]. The mode selector based on counter-tapered coupler consists of two waveguides with width changing inversely. These two waveguides are long enough to ensure adiabatic mode conversion. If we want to convert the mode of the upper left waveguide into the mode of the lower waveguide and output from the right port, the effective refractive index curve of the two modes must have a cross-point along the direction of propagation. In the two-mode MDM with TE_0 and TE_1, the TE_0 mode can input from the upper waveguide as long as the effective refractive index curve of the upper waveguide TE_0 and the effective refractive index curve of the lower waveguide TE_1 have a cross-point in the direction of propagation, then the input TE_0 of the upper waveguide can eventually be coupled into the lower waveguide and the two-mode MDM is achieved.

Figure 3.28 shows a two-mode Si-MDM link with TE_0 and TE_1, which is composed of two counter-tapered couplers and connected through a multi-mode transmission waveguide in the middle. We designed and optimized the device in a SOI strip waveguide with 220 nm top silicon and SiO_2 upper and lower cladding. The mode multiplexer and demultiplexer of the link are identical according to the reciprocity of the light. The working principle of the mode multiplexer can refer to Ref. [155]. As the effective refractive index of the guided mode increases with the width of the waveguide and the width of the two waveguides changes oppositely, it is possible to have a cross-point of the effective refractive index curve of the two waveguides as long as the width on the two sides of the waveguide is reasonably designed. As a result, the mode conversion may occur. The analytic method can also be used to analyze the problem from the perspective of local normal mode. The details can refer to Ref. [154].

Here, in addition to a counter-tapered coupler in the structure of the whole mode multiplexer, there is a bend waveguide on both sides to remove the coupling between the ports, and can also reduce the radiation loss caused by cutting off the waveguide suddenly. The width of the tapered waveguide (called WG-1) above the coupler is reduced from 0.49 μm to 0.24 μm, while the width of the waveguide on the bottom of the coupler (called WG-0) is increased from 0.61 μm to 0.86 μm. The length of the coupler is L_{tp}. The gap of the left input waveguides is 3 μm, and the upper waveguide is connected to WG-1 through a S-bend waveguide. The gap between the multimode waveguides in the upper waveguide on the right output section is slowly increased through a curved waveguide with a radius of 250 μm. The length of the left input part and the right output part are 20 μm and 40 μm, respectively.

We use the two-dimensional simulation software FIMMWAVE to calculate the relationship between the effective refractive index and the width of the two waveguide modes in the counter-tapered coupler when the wavelength is 1.55 μm, as shown in Fig. 3.29. We found that the TE_1 curve of WG-0 has a cross-point with the TE_0 curve of WG-1, which shows that when L_{tp} is long enough to ensure that the width of the waveguide changes slowly enough, the TE_0 of WG-1 can be adiabatically converted to TE_1 of WG-0. For the TE_0 curve of WG-0, it has no cross-point with

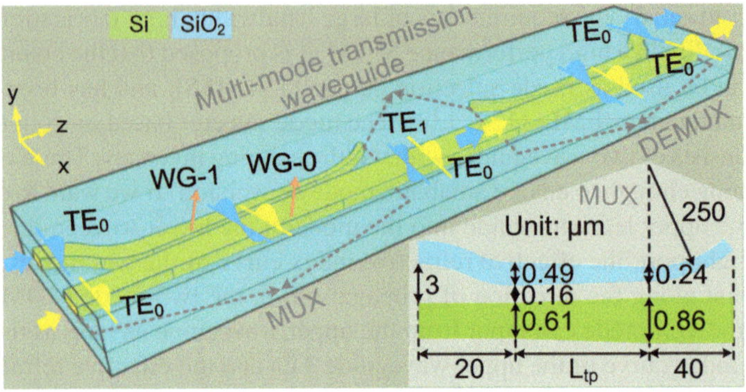

Fig. 3.28 Schematic of the two-mode MDM link based on two counter-tapered couplers. Inset: the design parameters for the multiplexer. Reproduced from Ref. [156] by permission of Optical Society of America

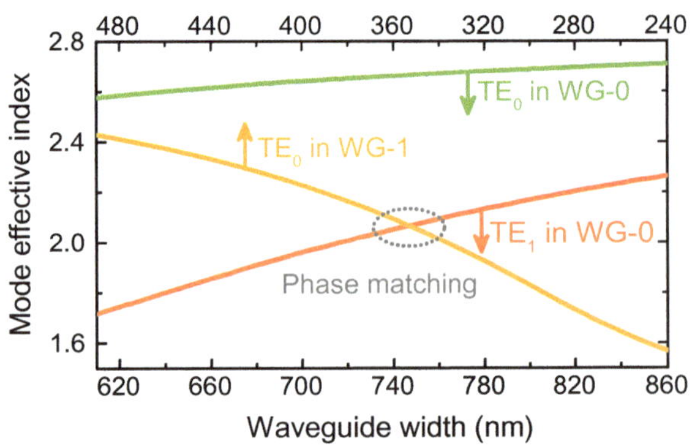

Fig. 3.29 Effective refractive index of the modes in each waveguide of the coupler with width tapering from 490 nm/610 nm to 240 nm/860 nm. Reproduced from Ref. [156] by permission of Optical Society of America

the effective refractive index curve of other guided modes, so the TE_0 input mode from WG-0 will not be converted to other modes, but propagating directly in the WG-0 waveguide. The width difference between the tapered waveguides is 0.25 μm, which ensures that the effective refractive index difference is large enough to avoid coupling optical power to the input waveguide again. Smaller waveguide width difference may cause unnecessary mode interference or radiation, and larger width difference may increase device length. After the TE_0 and TE_1 modes are multiplexed in the counter-tapered coupler, they will transmit in the multimode waveguide and are demultiplexed to different output ports in the next de-multiplexer.

In order to evaluate the performances of the mode multiplexer, we used a three-dimensional simulation software FIMMPROP base on EME to calculate the mode conversion loss. The device parameters are shown in the internal diagram of Fig. 3.28. Figure 3.30a shows the mode conversion loss from the TE_0 mode in the upper left port (called CH1) to the TE_1 mode in the lower right multimode waveguide when the wavelength is 1.55 μm. The mode conversion loss in the mode multiplexer

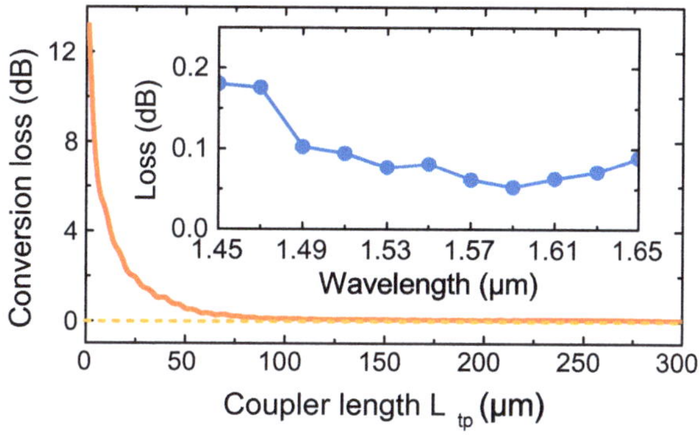

Fig. 3.30 Mode conversion from TE_0 in CH1 to TE_1 in the multi-mode waveguide. Inset: wavelength dependence of the mode conversion loss at $L_{tp} = 200\,\mu m$. Reproduced from Ref. [156] by permission of Optical Society of America

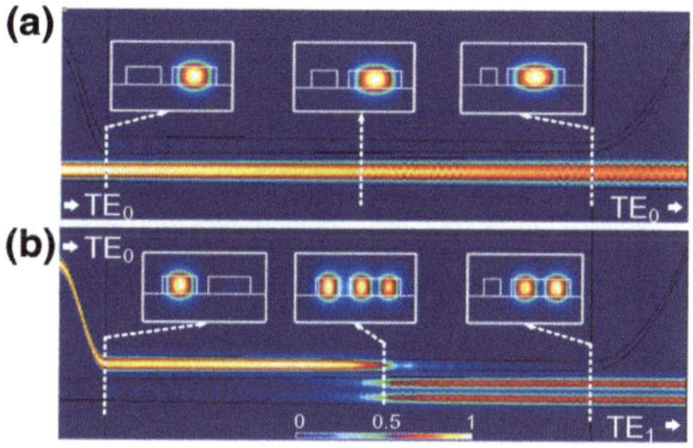

Fig. 3.31 **a–b** Mode propagation in the multiplexer for the TE_0 input from CH0 and CH1, respectively. Reproduced from Ref. [156] by permission of Optical Society of America

decreases with the increase of coupler length L_{tp}. When L_{tp} is larger than 120 μm, a very low loss of less than 0.1 dB can be obtained. Here, we choose $L_{tp} = 200\,\mu m$ to guarantee good performances, and a relatively small device size simultaneously. Moreover, the mode multiplexer also shows good broadband characteristics, and the mode conversion loss remains less than 0.18 dB when the wavelength is between 1.45 μm to 1.65 μm.

Figure 3.31a–b show the propagation of the TE_0 mode in the mode multiplexer when TE_0 inputs from the bottom waveguide (defined as CH0) or CH1. These figures show that the mode conversion in the mode multiplexer is just as expected. When TE_0 inputs from CH0, it will pass directly through the whole mode multiplexer, and there is almost no power transferred to other waveguides or modes. When TE_0 inputs from CH1, the mode power will slowly evolve into the bottom waveguide and output as the TE_1 mode.

We fabricated these MDM devices on SOI wafers with 220 nm top silicon layer and 2 μm buried oxide layer. First, we spin 100 nm HSQ on the SOI wafer as negative photoresist for electron beam lithography. Secondly, the pattern is defined by electron beam lithography. Then, the silicon layer is etched by dry etching. Finally, a 2 μm SiO_2 upper layer is deposited. Figure 3.32 shows the microscope picture of a two-mode Si-MDM link and its reference waveguide, and locally enlarges the input and output parts of the mode multiplexer.

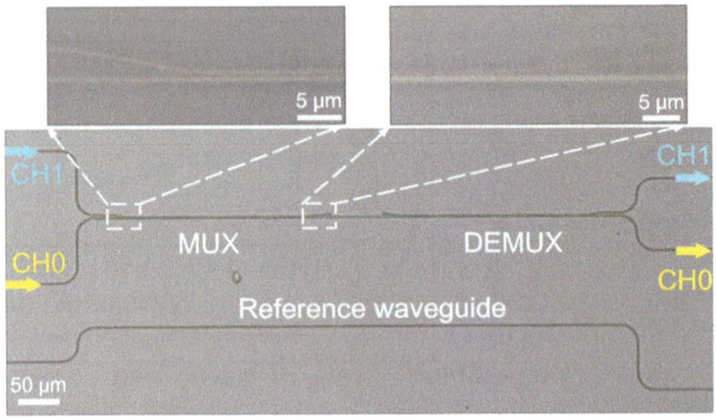

Fig. 3.32 Microscope photos of the two-mode MDM link with a 60 μm long multi-mode waveguide. Reproduced from Ref. [156] by permission of Optical Society of America

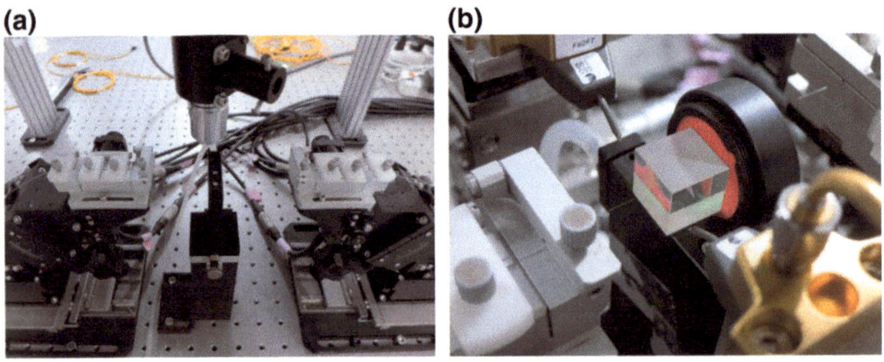

Fig. 3.33 **a** Butt-coupling test set-up. **b** Polarization calibration of input light by using PBS crystal

We characterized these device samples in a butt-coupling test platform, as shown in Fig. 3.33a. We used a tunable laser (Agilent 8164B, 81600-201) as the light source and a tricyclic polarization controller to control the polarization state. The light is coupled through a lensed fiber into the device and the output light passes through a lensed fiber and finally detected in the power-meter (Agilent 81624B). Before measuring these devices, we first calibrate the polarization state of the input light by using a broadband polarization beam crystal (GCC-402131) and a handhold optical powermeter (Thorlabs PM122D) with a probe to ensure that the extinction ratio of the TE to TM is ~20 dB, as shown in Fig. 3.33b.

Figure 3.34a shows the transmission spectrum of our designed two-mode Si-MDM link. Here, we define T_{ij} as the transmission characteristics from port CHi to port CHj. When the wavelength is 1550 nm, the excess loss of TE_0 through path (-T_{00}) and TE_1 through path (-T_{11}) are 0.41 dB and 0.74 dB, respectively. The reason why T_{11} is slightly smaller than T_{00} is that the TE_1 through path requires two mode conversion, and the TE_1 through path is similar to a long waveguide that does not have any mode conversion. Moreover, the bandwidth of T_{00} is larger than that of the

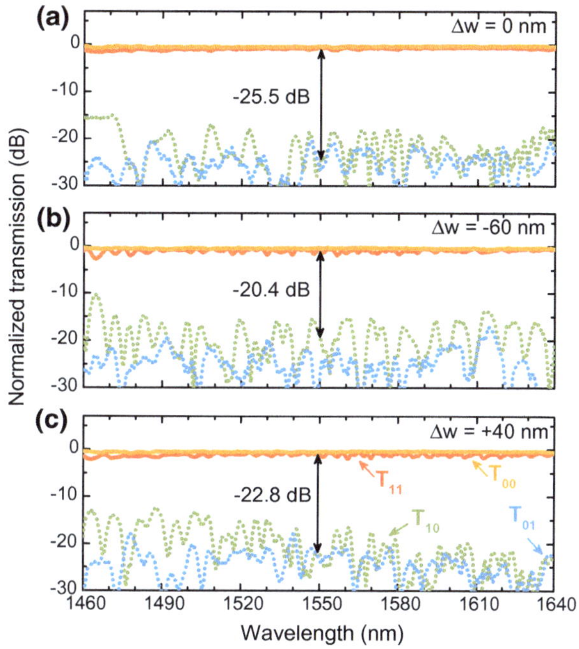

Fig. 3.34 Normalized transmission characteristics of the MDM link with **a** designed coupler sizes, and the width deviations of **b** $\Delta w = -60$ nm and **c** $\Delta w = 40$ nm. Reproduced from Ref. [156] by permission of Optical Society of America

T_{11}, and the loss variation is only 0.5 dB in the wavelength range from 1460 nm to 1640 nm. Nevertheless, T_{11} still has –1 dB bandwidth of more than 180 nm, which exceeds the tunable range of our laser wavelength. For input from CH1 or CH0, the crosstalk of the entire two-mode Si-MDM at a wavelength of 1550 nm is very low, only –25.5 and –26.1 dB. The crosstalk performance decreases at other wavelengths, probably due to the jitter of the polarization state of the tunable laser [140]. Even so, the crosstalk of the two modes is less than –13 dB in the whole 180 nm wavelength range.

In order to analyze the fabrication tolerance of this MDM mechanism, we also fabricated another two samples with width deviation $\Delta w = -60$ nm and $\Delta w = 40$ nm, respectively. Figure 3.34b–c show the normalized transmission spectrum of the two samples, respectively. We find that performance is only slightly degraded in such a larger error range. The two samples show considerable bandwidth, which is mainly attributed to the broadband advantage of mode evolution. We can find some ripples on the T_{11} curve, which may be due to some unnecessary mode interference in the counter-tapered coupler. These two samples prove that the MDM based on counter-tapered coupler has a large fabrication tolerance, which is helpful to improve the fabrication yield.

3.3.3 Three-Mode Si-MDM Link Based on Mode-Evolution Counter-Tapered Couplers

In previous chapter, we have proved a two-mode Si-MDM link. According to the design theory of the counter-tapered coupler, we can easily extend it to the multiplexing of higher order mode, which is also an obvious advantage compared to other broadband MDM based on mode evolution effect.

Figure 3.35 shows the microscope picture of a three-mode Si-MDM sample. It works at three modes of $TE_0/TE_1/TE_2$. The output part is not shown in the diagram. This TE_0/TE_1 part of the three-mode Si-MDM link can be seen in the internal diagram of Fig. 3.28. The design of the TE_2 multiplexing part can be done by calculating the effective refractive index curve of the TE_0 of the upper waveguide and the effective refractive index curve of the TE_2 of the lower waveguide. We find that when the width of the upper waveguide is reduced from 0.4 μm to 0.2 μm, and the width of the lower waveguide is reduced from 1 μm to 1.2 μm, the refractive index curve will have a cross-point in the propagation direction of the coupler. This will lead to that the TE_0 mode in the upper waveguide can be converted to the TE_2 mode in the lower waveguide. We also choose 200 μm as the length of the TE_2 mode coupler.

Figure 3.36a–c show the transmission characteristics of the three-mode MDM link when the TE_0 mode inputs from CH0, CH1 and CH2, respectively. When the wavelength is 1550 nm, the crosstalk are –20.5 dB, –15.9 dB and –14.4 dB, respectively. Although the performance decreases slightly as the number of multiplexed mode increases, the crosstalk performance can remain less than or equal to –10 dB from 1460 nm to 1640 nm wavelength. Such broadband properties also help to achieve broadband hybrid multiplexing with WDM or/and PDM. Moreover, we believe that when we further optimize the mode multiplexer, the MDM mechanism can be more easily extended to higher order mode.

In our proposed MDM mechanism, in addition to the mode conversion loss, the propagation loss of the waveguide is also an important source of total loss, especially when the device is relatively long. The propagation loss of waveguide can be improved by optimizing the process. Some relative work can refer to the paper reported by our research group previously [127]. Moreover, because our device is tested on a butt-coupling platform, the Fabry-Perot reflection between the optical

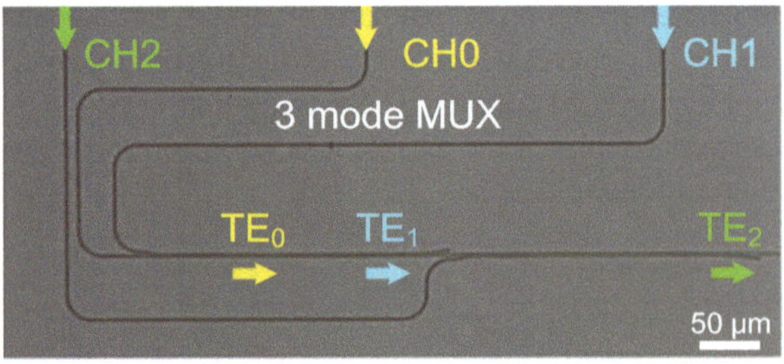

Fig. 3.35 Microscoped photos of the fabricated three-mode MDM. Reproduced from Ref. [156] by permission of Optical Society of America

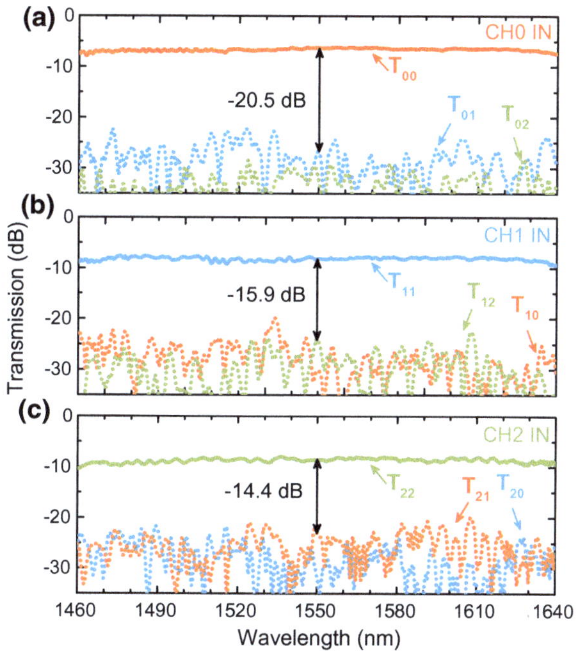

Fig. 3.36 **a–c** Transmission characteristics of the three-mode MDM link when the input channels are CH0, CH1, and CH2, respectively. Reproduced from Ref. [156] by permission of Optical Society of America

fiber and the waveguide will cause some unnecessary input light, which also reduce the polarization extinction ratio of the input light. This will eventually result in a relatively high crosstalk performance of the MDM link. One possible solution is to use grating coupling rather than butt coupling. Here the grating coupler is equivalent to a polarization filter that can improve the crosstalk performance [141]. Of course, the grating coupler is bandwidth-limited itself, which has an impact on the bandwidth performance of the MDM.

We also calculated the effect of temperature on mode conversion efficiency. When the temperature changes from −25 °C to 75 °C, the refractive index of Si changes from 3.4678 to 3.4861 at a wavelength of 1550 nm. However, the relative variation of SiO_2 in the upper and lower cladding layers is also similar. Therefore, the change of the refractive index does not obviously change the cross-point of the TE_0 and TE_1 curves in Fig. 3.29, so that the mode conversion efficiency remains high enough when the mode coupler is long enough. Through simulation, we find that the mode conversion efficiency from the TE_0 mode in the upper waveguide to the TE_1 mode in the lower waveguide changes by only less than 5%.

Table 3.6 Performance summary of our Si-AWG

	Gaussian	Gaussian	Gaussian	Flat-top
Channel spacing ($\Delta\lambda_{CH}$) (nm)	0.8	1.6	3.2	3.2
Channel number (N_{CH})	8	8	8	8
Size (μm^2)	260×1780	260×922	260×500	140×223
Insertion loss (IL) (dB)	2.92	2.90	2.32	3.60
Non-uniformity (L_u) (dB)	2.63	2.60	2.34	2.34
Crosstalk (XT) (dB)	−16.9 ~ −17.8	−17 ~ −19.3	−20.5 ~ −24.5	~ −15

3.4 Conclusion

In this chapter, we have further expanded three types of multiplexing scheme for silicon-based optical interconnections, and proposed or implemented a series of novel multiplexing devices, including

1. In the research of WDM devices, we have successfully implemented high performance Gaussian-shaped Si-AWG devices by using commercial 0.13 μm CMOS technology, with channel spacing of 0.8 nm, 1.6 nm, and 3.2 nm. We take 3.2 nm Si-AWG as an example. Through comparative study, we find that a comprehensive optimization method of broadening the arrayed waveguide and introducing a shallowly-etching mode converter near FPR can improve the performance of the device more effectively than a single optimization method. Based on this, we successfully implemented a Si-AWG router with channel spacing of 0.8 nm with a stable CMOS technology for DWDM applications. Rotating cycle test results show that the two devices work well. We further test the transmission characteristics of the two AWG in a 3-channel high-speed WDM system, including the bit error rate and the eye diagram of the multiplexed signal. Finally, we also demonstrate the Si-AWG devices with the other channel spacing and the flat-top Si-AWG preliminarily. The performance summary of these AWG devices is shown in Table 3.6.
2. In the research of PDM device, we first proposed a scheme to achieve ultra-bandwidth PBS by using the mode-sorting effect of asymmetric Y-junction. We achieve different mode-sorting functions by cascading three Y-junctions and carefully optimizing their design parameters. Finally, the overall device exhibits excellent performance in the wavelength range from 1450 nm to 1750 nm. We further discuss the effect of typical process error on the performance, and put forward some feasible schemes for further performance optimization.
3. In the aspect of MDM device research, we use counter-tapered couplers to realize a MDM link on silicon, which has the advantages of broadband, high fabrication tolerance and scalability. We analyze and optimize the mode multiplexer based on counter-tapered coupler. The test results show that for a two-mode MDM link, the –1 dB bandwidth is larger than 180 nm, and crosstalk is less than –13 dB in the

wavelength range of 1460–1640 nm. The performance of this two-mode MDM link is stable for fabrication errors (device width deviation changes from –60 nm to 40 nm) and temperature changes (–25°C to 75°C). We further expand it to the three-mode MDM link, with less than –10 dB crosstalk at 180 nm wavelength range.

References

1. Bozinovic N, Yue Y, Ren Y, Tur M, Kristensen P, Huang H, Willner AE, Ramachandran S (2013) Terabit-scale orbital angular momentum mode division multiplexing in fibers. Science 340:1545–1548
2. Wang J, Yang JY, Fazal IM, Ahmed N, Yan Y, Huang H, Ren Y, Yue Y, Dolinar S, Tur M, Willner AE (2012) Terabit free-space data transmission employing orbital angular momentum multiplexing. Nat Photonics 6:488–496
3. Smit MK (1988) New focusing and dispersive planar component based on an optical phased array. Electron Lett 24:385–386
4. Adar R, Henry CH, Dragone C, Kistler RC, Milbrodt MA (1993) Broad-band array multiplexers made with silica waveguides on silicon. J Lightw Technol 11:212–219
5. Malik A, Muneeb M, Pathak S, Shimura Y, Van Campenhout J, Loo R, Roelkens G (2013) Broad-band array multiplexers made with silica waveguides on silicon. IEEE Photonics Technol Lett 25:1805–1808
6. Fukazawa T, Ohno F, Baba T (2004) Very compact arrayed-waveguide-grating demultiplexer using Si photonic wire waveguides. Jpn J Appl Phys 43:L673
7. Zirngibl M, Dragone C, Joyner CH (1992) Demonstration of a 15×15 arrayed waveguide multiplexer on InP. IEEE Photonics Technol Lett 4:1250–1253
8. Yang B, Zhu Y, Jiao Y, Yang L, Sheng Z, He S, Dai D (2011) Compact arrayed waveguide grating devices based on small SU-8 strip waveguides. J Lightw Technol 29:2009–2014
9. Okayama H, Kawahara M, Kamijoh T (1996) Reflective waveguide array demultiplexer in $LiNbO_3$. J Lightw Technol 14:985–990
10. Liu WJ, Lai YC, Weng MH, Chen CM, Lee PH (2005) Simulation and fabrication of silicon oxynitride array waveguide grating for optical communication. In: Optical components and materials II, pp 43–54
11. Dai D, Wang Z, Bauters JF, Tien MC, Heck MJ, Blumenthal DJ, Bowers JE (2011) Low-loss Si_3N_4 arrayed-waveguide grating (de)multiplexer using nano-core optical waveguides. Opt Express 19:14130–14136
12. Muneeb M, Chen X, Verheyen P, Lepage G, Pathak S, Ryckeboer E, Malik A, Kuyken B, Nedeljkovic M, Van Campenhout J, Mashanovich GZ (2013) Demonstration of silicon-on-insulator mid-infrared spectrometers operating at 3.8 μm. Opt Express 21:11659–11669
13. Sasaki K, Ohno F, Motegi A, Baba T (2005) Arrayed waveguide grating of 70×60 μm^2 size based on Si photonic wire waveguides. Electron Lett 41:801–802
14. Ohno F, Sasaki K, Motegi A, Baba T (2006) Reduction in sidelobe level in ultracompact arrayed waveguide grating demultiplexer based on Si wire waveguide. Jpn J Appl Phys 45:6126
15. Dumon P, Bogaerts W, Van Thourhout D, Taillaert D, Wiaux V, Beckx S, Wouters J, Baets R (2004) Wavelength-selective components in SOI photonic wires fabricated with deep UV lithography. In: IEEE 1st international conference on group IV photonics, pp 28–30
16. Dumon P, Bogaerts W, Van Thourhout D, Taillaert D, Baets R, Wouters J, Beckx S, Jaenen P. (2006) Compact wavelength router based on a silicon-on-insulator arrayed waveguide grating pigtailed to a fiber array. Opt Express 14:664–669

17. Bogaerts W, Dumon P, Van Thourhout D, Taillaert D, Jaenen P, Wouters J, Beckx S, Wiaux V, Baets RG (2006) Compact wavelength-selective functions in silicon-on-insulator photonic wires. IEEE J Sel Top Quantum Electron 12:1394–1401
18. Bogaerts W, Selvaraja SK, Dumon P, Brouckaert J, De Vos K, Van Thourhout D, Baets R (2010) Silicon-on-insulator spectral filters fabricated with CMOS technology. IEEE J Sel Top Quantum Electron 16:33–44
19. Pathak S, Vanslembrouck M, Dumon P, Van Thourhout D, Bogaerts W (2011) Compact 16×16 channels routers based on silicon-on-insulator AWGs. In: 16th annual symposium of the IEEE photonics benelux chapter, pp 101–104
20. Wang L, Bogaerts W, Dumon P, Selvaraja SK, Teng J, Pathak S, Han X, Wang J, Jian X, Zhao M, Baets R (2012) Athermal arrayed waveguide gratings in silicon-on-insulator by overlaying a polymer cladding on narrowed arrayed waveguides. Appl Opt 51:1251–1256
21. Pathak S, Vanslembrouck M, Dumon P, Van Thourhout D, Bogaerts W (2012) Compact SOI-based polarization diversity wavelength de-multiplexer circuit using two symmetric AWGs. Opt Express 20:B493–B500
22. Pathak S, Vanslembrouck M, Dumon P, Van Thourhout D, Bogaerts W (2013) Optimized silicon AWG with flattened spectral response using an MMI aperture. J Lightw Technol 31:87–93
23. Pathak S, Van Thourhout D, Bogaerts W (2013) Design trade-offs for silicon-on-insulator-based AWGs for (de)multiplexer applications. Opt Lett 38:2961–2964
24. Pathak S, Vanslembrouck M, Dumon P, Van Thourhout D, Verheyen P, Lepage G, Absil P, Bogaerts W (2014) Effect of mask discretization on performance of silicon arrayed waveguide gratings. IEEE Photonics Technol Lett 26:718–721
25. Pathak S, Dumon P, Van Thourhout D, Bogaerts W (2014) Comparison of AWGs and echelle gratings for wavelength division multiplexing on silicon-on-insulator. IEEE Photonics J 6:4900109
26. Kim DJ, Lee JM, Song JH, Pyo J, Kim G (2008) Crosstalk reduction in a shallow-etched silicon nanowire AWG. IEEE Photonics Technol Lett 20:1615–1617
27. Kim DJ, Pyo J, Kim G (2009) Center wavelength uniformity of shallow-etched silicon photonic wire AWG. In: IEEE 6th international conference on group IV photonics, pp 128–130
28. Cheung ST, Guan B, Djordjevic SS, Okamoto K, Yoo SB (2012) Low-loss and high contrast silicon-on-insulator (SOI) arrayed waveguide grating. In: CLEO: science and innovations
29. Cheung S, Su T, Okamoto K, Yoo SJB (2014) Ultra-compact silicon photonic 512×512 25 GHz arrayed waveguide grating router. IEEE J Sel Top Quantum Electron 20:310–316
30. Okamoto K, Ishida K (2013) Fabrication of silicon reflection-type arrayed-waveguide gratings with distributed Bragg reflectors. Opt Lett 38:3530–3533
31. Okamoto K (2014) Wavelength-division-multiplexing devices in thin SOI: advances and prospects. IEEE J Sel Top Quantum Electron 20:248–257
32. Dai D, He S (2006) Ultrasmall overlapped arrayed-waveguide grating based on Si nanowire waveguides for dense wavelength division demultiplexing. IEEE J Sel Top Quantum Electron 12:1301–1305
33. Dai D, He S (2006) Novel ultracompact Si-nanowire-based arrayed-waveguide grating with microbends. Opt Express 14:5260–5265
34. Dai D, Liu L, Wosinski L, He SAHS (2006) Design and fabrication of ultra-small overlapped AWG demultiplexer based on α-Si nanowire waveguides. Electron Lett 42:400–402
35. Fu X, Dai D (2011) Ultra-small Si-nanowire-based 400 GHz-spacing 15×15 arrayed-waveguide grating router with microbends. Electron Lett 47:266–268
36. Chen S, Fu X, Wang J, Shi Y, He S, Dai D (2015) Compact dense wavelength-division (de)multiplexer utilizing a bidirectional arrayed-waveguide grating integrated with a Mach-Zehnder interferometer. J Lightw Technol 33:2279–2285
37. Chen S, Shi Y, He S, Dai D (2015) Compact monolithically-integrated hybrid (de) multiplexer based on silicon-on-insulator nanowires for PDM-WDM systems. Opt Express 23:12840–12849

38. Lang T, He JJ, Kuang JG, He S (2007) Birefringence compensated AWG demultiplexer with angled star couplers. Opt Express 15:15022–15028
39. Zou J, Lang T, Wang L, He JJ (2011) Uniform polarization-dispersion compensation of all channels in highly birefringent silicon nanowire-based arrayed waveguide grating. IEEE Photonics Technol Lett 23:1787–1789
40. Zou J, Jiang X, Xia X, Lang T, He JJ (2013) Ultra-compact birefringence-compensated arrayed waveguide grating triplexer based on silicon-on-insulator. J Lightw Technol 31:1935–1940
41. Zou J, Xia X, Chen G, Lang T, He JJ (2014) Birefringence compensated silicon nanowire arrayed waveguide grating for CWDM optical interconnects. Opt Lett 39:1834–1837
42. Ye T, Fu Y, Qiao L, Chu T (2014) Low-crosstalk Si arrayed waveguide grating with parabolic tapers. Opt Express 22:31899–31906
43. Zhang J, An J, Zhao L, Song S, Wang L, Li J, Wang H, Wu Y, Hu X (2011) Ultra-compact triplexing filters based on SOI nanowire AWGs. J Semicond 32:044009
44. Lei Z, Junming A, Jiashun Z, Shijiao S, Yuanda W, Xiongwei H (2011) 16 channel 200 GHz arrayed waveguide grating based on Si nanowire waveguides. J Semicond 32:024010
45. Zhao L, An J, Zhang J, Song S, Wu Y, Hu X (2010) Theoretical analysis of a novel polarization-insensitive AWG demultiplexer based on Si nanowire and slot waveguides. In: Optoelectronic devices and integration III, pp 78472A
46. Huang H, Ho ST, Huang D, Tu Y, Liu W (2010) Design of temperature-independent arrayed waveguide gratings based on the combination of multiple types of waveguide. Appl Opt 49:3025–3034
47. Li H, Li E, Liu Z, Wei K, Dong X, Bai Y (2012) Design of 1×8 silicon nanowire arrayed waveguide grating for on-chip arrayed waveguide grating demodulation integration microsystem. Opt Eng 51:123001
48. Li H, Bai Y, Dong X, Li E, Li Y, Liu Y, Zhou W (2013) Optimal design of an ultrasmall SOI-based 1×8 flat-top AWG by using an MMI. Sci World J
49. Li H, Bai Y, Dong X, Li E, Li Y, Zhou W, Liu Y (2013) Practical fabrication and analysis of an optimized compact eight-channel silicon arrayed-waveguide grating. Opt Eng 52:064602
50. Li H, Li Y, Li E, Dong X, Bai Y, Liu Y, Zhou W (2013) Temperature-insensitive arrayed waveguide grating demodulation technique for fiber Bragg grating sensors. Opt Laser Technol 51:77–81
51. Li H, Zhou W, Liu Y, Dong X, Zhang C, Miao C, Zhang M, Li E, Tang C (2014) Preliminary investigation of an SOI-based arrayed waveguide grating demodulation integration microsystem. Sci Rep 4:4848
52. Yebo NA, Bogaerts W, Hens Z, Baets R (2011) On-chip arrayed waveguide grating interrogated silicon-on-insulator microring resonator-based gas sensor. IEEE Photonics Technol Lett 23:1505–1507
53. Cheben P, Schmid JH, Delâge A, Densmore A, Janz S, Lamontagne B, Lapointe J, Post E, Waldron P, Xu DX (2007) A high-resolution silicon-on-insulator arrayed waveguide grating microspectrometer with sub-micrometer aperture waveguides. Opt Express 15:2299–2306
54. Cheben P, Post E, Janz S, Albert J, Laronche A, Schmid JH, Xu DX, Lamontagne B, Lapointe J, Delâge A, Densmore A (2008) Tilted fiber Bragg grating sensor interrogation system using a high-resolution silicon-on-insulator arrayed waveguide grating. Opt Lett 33:2647–2649
55. Dai D, Fu X, Shi Y, He S (2010) Experimental demonstration of an ultracompact Si-nanowire-based reflective arrayed-waveguide grating (de)multiplexer with photonic crystal reflectors. Opt Lett 35:2594–2596
56. Shi Y, Fu X, Dai D (2010) Design and fabrication of a 200 GHz Si-nanowire-based reflective arrayed-waveguide grating (de)multiplexer with optimized photonic crystal reflectors. Appl Opt 49:4859–4865
57. Dai D (2005) Modeling, design and fabrication of arrayed-waveguide gratings. PhD Disseration, Zhejiang University
58. Huang H (2010) The study of silicon photonic-wires based arrayed waveguide grating. PhD Disseration, Huazhong University of Science Technology

59. Dumon P (2007) Ultra-compact integrated optical filters in silicon-on-insulator by means of wafer-scale technology. PhD Disseration, Ghent University
60. Pathak S (2014) Silicon nano-photonics based arrayed waveguide gratings. PhD Disseration, Ghent University
61. Wang J, Sheng Z, Li L, Pang A, Wu A, Li W, Wang X, Zou S, Qi M, Gan F (2014) Low-loss and low-crosstalk 8×8 silicon nanowire AWG routers fabricated with CMOS technology. Opt Express 22:9395–9403
62. Sheng Z, Dai D, He S (2007) Improve channel uniformity of an Si-nanowire AWG demultiplexer by using dual-tapered auxiliary waveguides. J Lightw Technol 25:3001–3007
63. Gan F, Barwicz T, Popovic MA, Dahlem MS, Holzwarth CW, Rakich PT, Smith HI, Ippen EP, Kartner FX (2007) Maximizing the thermo-optic tuning range of silicon photonic structures. Photon Switch 67–68
64. Soole JBD, Amersfoort MR, LeBlanc HP, Andreadakis NC, Rajhel A, Caneau C, Bhat R, Koza MA, Youtsey C, Adesida I (1996) Use of multimode interference couplers to broaden the passband of wavelength-dispersive integrated WDM filters. IEEE Photonics Technol Lett 8:1340–1342
65. Dai D, Mei W, He S (2003) Using a tapered MMI to flatten the passband of an AWG. Opt Commun 219:233–239
66. Okamoto K, Sugita A (1996) Flat spectral response arrayed-waveguide grating multiplexer with parabolic waveguide horns. Electron Lett 32:1661
67. Kiyat I, Aydinli A, Dagli N (2005) A compact silicon-on-insulator polarization splitter. IEEE Photonics Technol Lett 17:100–102
68. Fukuda H, Yamada K, Tsuchizawa T, Watanabe T, Shinojima H, Itabashi SI (2006) Ultrasmall polarization splitter based on silicon wire waveguides. Opt Express 14:12401–12408
69. Komatsu MA, Saitoh K, Koshiba M (2009) Design of miniaturized silicon wire and slot waveguide polarization splitter based on a resonant tunneling. Opt Express 17:19225–19233
70. Yue Y, Zhang L, Yang JY, Beausoleil RG, Willner AE (2010) Silicon-on-insulator polarization splitter using two horizontally slotted waveguides. Opt Lett 35:1364–1366
71. Lin S, Hu J, Crozier KB (2011) Ultracompact, broadband slot waveguide polarization splitter. Appl Phys Lett 98:151101
72. Zhang H, Huang Y, Das S, Li C, Yu M, Lo PG, Hong M, Thong J (2013) Polarization splitter using horizontal slot waveguide. Opt Express 21:3363–3369
73. Kim JT (2014) CMOS-compatible polarization splitter for 3-D silicon photonic integrated circuits. J Lightw Technol 32:2123–2127
74. Kim DW, Lee MH, Kim Y, Kim KH (2015) Planar-type polarization beam splitter based on a bridged silicon waveguide coupler. Opt Express 23:998–1004
75. Ying Z, Wang G, Zhang X, Ho HP, Huang Y (2015) Ultracompact and broadband polarization beam splitter based on polarization-dependent critical guiding condition. Opt Lett 40:2134–2137
76. Rahman BMA, Somasiri N, Themistos C, Grattan KTV (2001) Design of optical polarization splitters in a single-section deeply etched MMI waveguide. Appl Phys B 73:613–618
77. Hong JM, Ryu HH, Park SR, Jeong JW, Lee SG, Lee EH, Park SG, Woo D, Kim S, Beom-Hoan O (2003) Design and fabrication of a significantly shortened multimode interference coupler for polarization splitter application. IEEE Photonics Technol Lett 15:72–74
78. Shi Y, Dai D, He S (2007) Proposal for an ultracompact polarization-beam splitter based on a photonic-crystal-assisted multimode interference coupler. IEEE Photonics Technol Lett 19:825–827
79. Shi Y, Shahid N, Li M, Berrier A, He S, Anand S (2010) Experimental demonstration of an ultracompact polarization beamsplitter based on a multimode interference coupler with internal photonic crystals. Opt Eng 49:060503
80. Jiao Y, Dai D, Shi Y, He S (2009) Shortened polarization beam splitters with two cascaded multimode interference sections. IEEE Photonics Technol Lett 21:1538–1540
81. Yang BK, Shin SY, Zhang D (2009) Ultrashort polarization splitter using two-mode interference in silicon photonic wires. IEEE Photonics Technol Lett 21:432–434

82. Hosseini A, Rahimi S, Xu X, Kwong D, Covey J, Chen RT (2011) Ultracompact and fabrication-tolerant integrated polarization splitter. Opt Lett 36:4047–4049
83. Huang Y, Tu Z, Yi H, Li Y, Wang X, Hu W (2013) High extinction ratio polarization beam splitter with multimode interference coupler on SOI. Opt Commun 3017:46–49
84. Xu Y, Xiao J, Sun X (2014) Compact polarization beam splitter for silicon-based slot waveguides using an asymmetrical multimode waveguide. J Lightw Technol 32:4282–4288
85. Yin M, Yang W, Li Y, Wang X, Li H (2015) CMOS-compatible and fabrication-tolerant MMI-based polarization beam splitter. Opt Commun 335:48–52
86. Rajarajan M, Themistos C, Rahman BA, Grattan KT (1997) Characterization of metal-clad TE/TM mode splitters using the finite element method. J Lightw Technol 15:2264–2269
87. Albrecht P, Hamacher M, Heidrich H, Hoffmann D, Nolting H, Weinert CM (1990) TE/TM mode splitters on InGaAsP/InP. IEEE Photonics Technol Lett 2:114–115
88. Okuno M, Sugita A, Jinguji K, Kawachi M (1994) Birefringence control of silica waveguides on Si and its application to a polarization-beam splitter/switch. IEEE Photonics Technol Lett 12:625–633
89. Wei PK, Wang WS (1994) A TE-TM mode splitter on lithium niobate using Ti, Ni, and MgO diffusions. IEEE Photonics Technol Lett 6:245–248
90. Van der Tol JJ, Laarhuis JH (1991) A polarization splitter on $LiNbO_3$ using only titanium diffusion. J Lightw Technol 9:879–886
91. Ghirardi F, Brandon J, Carre M, Bruno A, Menigaux L, Carenco A (1993) Polarization splitter based on modal birefringence in InP/InGaAsP optical waveguides. IEEE Photonics Technol Lett 5:1047–1049
92. Wang Q, Farrell G, Semenova Y (2006) Design of integrated polarization beam splitter with liquid crystal. IEEE J Sel Top Quantum Electron 12:1349–1353
93. Lin KC, Chuang WC, Lee WY (1996) Proposal and analysis of an ultrashort directional-coupler polarization splitter with an NLC coupling layer. J Lightw Technol 14:2547–2553
94. Liang TK, Tsang HK (2005) Integrated polarization beam splitter in high index contrast silicon-on-insulator waveguides. IEEE Photonics Technol Lett 17:393–395
95. Dai D, Wang Z, Bowers JE (2011) Considerations for the design of asymmetrical mach-zehnder interferometers used as polarization beam splitters on a submicrometer silicon-on-insulator platform. J Lightw Technol 29:1808–1817
96. Dai D, Wang Z, Peters J, Bowers JE (2012) Compact polarization beam splitter using an asymmetrical mach-zehnder interferometer based on silicon-on-insulator waveguides. IEEE Photonics Technol Lett 24:673–675
97. Kim S, Nordin GP, Cai J, Jiang J (2003) Ultracompact high-efficiency polarizing beam splitter with a hybrid photonic crystal and conventional waveguide structure. Opt Lett 28:2384–2386
98. Tang Y, Dai D, He S (2009) Proposal for a grating waveguide serving as both a polarization splitter and an efficient coupler for silicon-on-insulator nanophotonic circuits. IEEE Photonics Technol Lett 21:242–244
99. Wang Z, Tang Y, Wosinski L, He S (2010) Experimental demonstration of a high efficiency polarization splitter based on a one-dimensional grating with a Bragg reflector underneath. IEEE Photonics Technol Lett 22:1568–1570
100. Zaoui WS, Kunze A, Vogel W, Berroth M (2013) CMOS-compatible polarization splitting grating couplers with a backside metal mirror. IEEE Photonics Technol Lett 25:1395–1397
101. Streshinsky M, Shi R, Novack A, Cher RT, Lim AE, Lo PG, Baehr-Jones T, Hochberg M (2013) A compact bi-wavelength polarization splitting grating coupler fabricated in a 220 nm SOI platform. Opt Express 21:31019–31028
102. Zou CL, Sun FW, Dong CH, Ren XF, Cui JM, Chen XD, Han ZF, Guo GC (2011) Broadband integrated polarization beam splitter with surface plasmon. Opt Lett 36:3630–3632
103. Chee J, Zhu S, Lo GQ (2012) CMOS compatible polarization splitter using hybrid plasmonic waveguide. Opt Express 20:25345–25355
104. Gao L, Hu F, Wang X, Tang L, Zhou Z (2013) Ultracompact and silicon-on-insulator-compatible polarization splitter based on asymmetric plasmonic-dielectric coupling. Appl Phys B 113:199–203

105. Tan Q, Huang X, Zhou W, Yang K (2013) A plasmonic based ultracompact polarization beam splitter on silicon-on-insulator waveguides. Sci Rep 3:2206
106. Kim S, Qi M (2014) Copper nanorod array assisted silicon waveguide polarization beam splitter. Opt Express 22:9508–9516
107. Ma Y, Farrell G, Semenova Y, Chan HP, Zhang H, Wu Q (2014) Low-loss, high extinction ration and ultra-compact plasmonic polarization beam splitter. IEEE Photonics Technol Lett 26:660–663
108. Xu Y, Xiao J, Sun X (2015) Proposal for compact polarization splitter using asymmetrical three-Guide directional coupler. IEEE Photonics Technol Lett 27:654–657
109. Dai D, Wang Z, Bowers JE (2011) Ultrashort broadband polarization beam splitter based on an asymmetrical directional coupler. Opt Lett 36:2590–2592
110. Dai D, Bowers JE (2011) Novel ultra-short and ultra-broadband polarization beam splitter based on a bent directional coupler. Opt Express 19:18614–18620
111. Wang J, Liang D, Tang Y, Dai D, Bowers JE (2013) Realization of an ultra-short silicon polarization beam splitter with an asymmetrical bent directional coupler. Opt Lett 38:4–6
112. Dai D (2012) Silicon polarization beam splitter based on an asymmetrical evanescent coupling system with three optical waveguides. J Lightw Technol 30:3281–3287
113. Wang J, Dai D (2013) Ultra-small silicon polarization beam splitter based on cascaded asymmetry directional couplers. In: Optical fiber communication conference
114. Lou F, Dai D, Wosinski L (2012) Ultracompact polarization beam splitter based on a dielectric-hybrid plasmonic-dielectric coupler. Opt Lett 37:3372–3374
115. Guan X, Wu H, Shi Y, Wosinski L, Dai D (2013) Ultracompact and broadband polarization beam splitter utilizing the evanescent coupling between a hybrid plasmonic waveguide and a silicon nanowire. Opt Lett 38:3005–3008
116. Guan X, Wu H, Shi Y, Dai D (2014) Extremely small polarization beam splitter based on a multimode interference coupler with a silicon hybrid plasmonic waveguide. Opt Lett 39:259–262
117. Watts MR, Haus HA, Ippen EP (2005) Integrated mode-evolution-based polarization splitter. Opt Lett 30:967–969
118. Su Z, Timurdogan E, Hosseini ES, Sun J, Leake G, Coolbaugh DD, Watts MR (2014) Four-port integrated polarizing beam splitter. Opt Lett 39:965–968
119. Uematsu T, Kitayama T, Ishizaka Y, Saitoh K (2014) Ultra-broadband silicon-wire polarization beam combiner/splitter based on a wavelength insensitive coupler with a point-symmetrical configuration. IEEE Photonics J 6:4500108
120. Love JD, Riesen N (2012) Single-, few-, and multimode Y-junctions. J Lightw Technol 30:304–309
121. Riesen N, Love JD (2012) Design of mode-sorting asymmetric Y-junctions. Appl Opt 51:2778–2783
122. Driscoll JB, Grote RR, Souhan B, Dadap JI, Lu M, Osgood RM (2013) Asymmetric Y junctions in silicon waveguides for on-chip mode-division multiplexing. Opt Lett 38:1854–1856
123. Driscoll JB, Grote RR., Souhan B, Dadap JI, Lu M, Osgood RM (2014) A 60 Gb/s MDM-WDM Si photonic link with < 0.7 dB power penalty per channel. Opt Express 22:18543–18555
124. Chen W, Wang P, Yang J (2013) Mode multi/demultiplexer based on cascaded asymmetric Y-junctions. Opt Express 21:25113–25119
125. Chen W, Wang P, Yang J (2014) Optical mode interleaver based on the asymmetric multimode Y junction. IEEE Photonics Technol Lett 26:2043–2046
126. Wang J, Qi M, Xuan Y, Huang H, Li Y, Li M, Chen X, Jia Q, Sheng Z, Wu A, Li W (2014) Ultra-broadband silicon-on-insulator polarization beam splitter based on cascaded mode-sorting asymmetric Y-junctions. IEEE Photonics J 6:1–8
127. Qiu C, Sheng Z, Li H, Liu W, Li L, Pang A, Wu A, Wang X, Zou S, Gan F (2014) Fabrication, characterization and loss analysis of silicon nanowaveguides. J Lightw Technol 32:2303–2307

128. Zhang Y, Yang S, Lim AEJ, Lo GQ, Galland C, Baehr-Jones T, Hochberg M (2013) A compact and low loss Y-junction for submicron silicon waveguide. Opt Express 21:1310–1316
129. Dai D, Bowers JE (2014) Silicon-based on-chip multiplexing technologies and devices for Peta-bit optical interconnects. Nanophotonics 3:283–311
130. Richardson DJ, Fini JM, Nelson LE (2013) Space-division multiplexing in optical fibres. Nat Photonics 7:354–362
131. Van Uden RG, Correa RA, Lopez EA, Huijskens FM, Xia C, Li G, Schülzgen A, De Waardt H, Koonen AM, Okonkwo CM (2014) Ultra-high-density spatial division multiplexing with a few-mode multicore fibre. Nat Photonics 8:865–870
132. Chen H, Sleiffer V, Snyder B, Kuschnerov M, van Uden R, Jung Y, Okonkwo CM, Raz O, O'Brien P, de Waardt H, Koonen T (2013) Demonstration of a photonic integrated mode coupler with MDM and WDM transmission. IEEE Photonics Technol Lett 25:2039–2042
133. Ding Y, Ou H, Xu J, Peucheret C (2013) Silicon photonic integrated circuit mode multiplexer. IEEE Photonics Technol Lett 25:648–651
134. Riesen N, Love JD (2013) Ultra-broadband tapered mode-selective couplers for few-mode optical fiber networks. IEEE Photonics Technol Lett 25:2501–2504
135. Chen H, Van Uden R, Okonkwo C, Koonen T (2014) Compact spatial multiplexers for mode division multiplexing. Opt Express 22:31582–31594
136. Gross S, Riesen N, Love JD, Withford MJ (2014) Three-dimensional ultra-broadband integrated tapered mode multiplexers. Laser Photonics Rev 8:L81–L85
137. Hanzawa N, Saitoh K, Sakamoto T, Matsui T, Tsujikawa K, Koshiba M, Yamamoto F (2014) Mode multi/demultiplexing with parallel waveguide for mode division multiplexed transmission. Opt Express 22:29321–29330
138. Riesen N, Gross S, Love JD, Withford MJ (2014) Femtosecond direct-written integrated mode couplers. Opt Express 22:29855–29861
139. Dai D, Wang J, Shi Y (2013) Silicon mode (de)multiplexer enabling high capacity photonic networks-on-chip with a single-wavelength-carrier light. Opt Lett 38:1422–1424
140. Wang J, He S, Dai D (2014) On-chip silicon 8-channel hybrid (de)multiplexer enabling simultaneous mode-and polarization-division-multiplexing. Laser Photonics Rev 8:L18–L22
141. Wang J, Chen P, Chen S, Shi Y, Dai D (2014) Improved 8-channel silicon mode demultiplexer with grating polarizers. Opt Express 22:12799–12807
142. Wang J, Chen S, Dai D (2014) Silicon hybrid demultiplexer with 64 channels for wavelength/mode-division multiplexed on-chip optical interconnects. Opt Lett 39:6993–6996
143. Dai D, Wang J, Chen S, Wang S, He S (2015) Monolithically integrated 64-channel silicon hybrid demultiplexer enabling simultaneous wavelength-and mode-division-multiplexing. Laser Photonics Rev 9:339–344
144. Luo LW, Ophir N, Chen CP, Gabrielli LH, Poitras CB, Bergmen K, Lipson M (2014) WDM-compatible mode-division multiplexing on a silicon chip. Nat Commun 5:1–7
145. Ding Y, Xu J, Da Ros F, Huang B, Ou H, Peucheret C (2013) On-chip two-mode division multiplexing using tapered directional coupler-based mode multiplexer and demultiplexer. Opt Express 21:10376–10382
146. Qiu H, Yu H, Hu T, Jiang G, Shao H, Yu P, Yang J, Jiang X (2013) Silicon mode multi/demultiplexer based on multimode grating-assisted couplers. Opt Express 21:17904–17911
147. Chen GFR, Wang T, Ooi KJA, Chee AKL, Ang LK, Tan DTH (2015) Wavelength selective mode division multiplexing on a silicon chip. Opt Express 23:8095–8103
148. Uematsu T, Ishizaka Y, Kawaguchi Y, Saitoh K, Koshiba M (2012) Design of a compact two-mode multi/demultiplexer consisting of multimode interference waveguides and a wavelength-insensitive phase shifter for mode-division multiplexing transmission. J Lightw Technol 30:2421–2426
149. Li Y, Li C, Li C, Cheng B, Xue C (2014) Compact two-mode (de)multiplexer based on symmetric Y-junction and Multimode interference waveguides. Opt Express 22:5781–5786
150. Ye M, Yu Y, Zou J, Yang W, Zhang X (2014) On-chip multiplexing conversion between wavelength division multiplexing-polarization division multiplexing and wavelength division multiplexing-mode division multiplexing. Opt Lett 39:758–761

151. Xing J, Li Z, Xiao X, Yu J, Yu Y (2013) Two-mode multiplexer and demultiplexer based on adiabatic couplers. Opt Lett 38:3468–3470
152. Dorin BA, Winnie NY (2014) Two-mode division multiplexing in a silicon-on-insulator ring resonator. Opt Express 22:4547–4558
153. Dai D (2014) Multimode optical waveguide enabling microbends with low inter-mode crosstalk for mode-multiplexed optical interconnects. Opt Express 22:27524–27534
154. Milton AF, Burns WK (1975) Tapered velocity couplers for integrated optics: design. Appl Opt 14:1207–1212
155. Riesen N, Love JD (2013) Tapered velocity mode-selective couplers. J Lightw Technol 31:2163–2169
156. Wang J, Xuan Y, Qi M, Huang H, Li Y, Li M, Chen X, Sheng Z, Wu A, Li W, Wang X (2015) Broadband and fabrication-tolerant on-chip scalable mode-division multiplexing based on mode-evolution counter-tapered couplers. Opt Lett 40:1956–1959

Chapter 4
CMOS-Compatible Si-Based Polarization Splitter-Rotator

4.1 Research Progress of Si-PSR

In the past, the research on silicon-based optical interconnections was mainly focused to implement various optical functional devices on silicon, such as silicon-based electrically-pumped laser, electro-optic modulator, photo-detector, WDM device and MDM device. In addition to on-chip optical interconnections, other kinds of optical interconnections inevitably need to connect to the outside world. Now the fiber is often used as an medium for external connection. However, on one hand, the polarization state in the optical fiber is random; on the other hand, the SOI waveguide has a larger material refractive index than the traditional integrated optical waveguide, which makes the effective refractive index of the TE and TM modes very different, causing the device performance extremely sensitive to the polarization state. Therefore, if the polarization sensitivity of device performance is not properly solved, silicon photonics will be limited without connecting to the outside world. It can not achieve more complicated device circuits or networks like traditional integrated optics, and can not achieve the goal of using optical interconnections instead of electrical interconnections. One solution is to specifically design their polarization-insensitive structures for each device [1, 2]. However, the devices with the polarization-insensitive design are generally not of the best performance, and these devices often require special device structures and complex process control. As a result, the polarization-insensitivity effect is difficult to guarantee. Another solution is to use the square waveguides, but this scheme requires precise control of the size, which is difficult to achieve in the process. In addition, performances in the coupling region and bend region are still polarization sensitive [3].

In 2007, researchers from MIT proposed a polarization diversity mechanism to solve the polarization sensitivity of silicon-based devices [4]. Light with random polarization coupled into the chip from the optical fiber is divided into two orthogonal components (generally the quasi-TE mode and the quasi-TM mode) to enter two separated waveguides.

J. Wang, *CMOS-Compatible Key Engineering Devices for High-Speed Silicon-Based Optical Interconnections*, Springer Theses,
https://doi.org/10.1007/978-981-13-3378-1_4

The quasi-TM mode input is converted to a quasi-TE mode on one of the waveguides by a polarization rotator (PR), and the input quasi-TE mode does not undergo any processing. The two beams pass through silicon based functional devices working on quasi TE mode, and realize various functions and signal processing. The output light is recombined into the original polarization state through the opposite processing, and is received by another optical fiber at the output end. Under such a mechanism, all functional devices work in the quasi-TE mode, and the external polarization state does not affect the internal operation. Through this mechanism, the design requirements on functional devices are greatly reduced, and the feasibility and application prospects of silicon-based photonic devices in optical interconnections and optical communications are greatly improved. Although this scheme is verified under the SiN platform, such a mechanism also provides reference to the SOI platform with higher refractive index difference.

In the polarization diversity scheme, there are two key components. One is PBS which is used to decompose the input polarization state into two orthogonal polarization states (PBS). Another one is PR which is used to convert the polarization state into another polarization state. For the state-of-art PBS, we have made a detailed review in Chap. 3.

For another important polarization-controlling device, i. e., PR, it can be divided into two types according to the mechanism of mode conversion. The first type of PR has relatively single structure, generally composed of two layers with the upper layer slowly narrowed and the lower layer gradually widened. It is like the waveguide was changed by 90°, and the polarization state was also rotated by 90°. In recent years, such PR devices based on SiN [5], SiN/SOI [6], all SOI [7] and hybrid plasmonic waveguide (HPW) [8] have been reported. The second type of PR, by breaking the symmetry of the cross-section of the waveguide, make the waveguide mode converted from the orthogonal mode such as TE and TM to a hybrid mode, so that the power between the two modes can be exchanged periodically. In order to break the symmetry of the waveguide cross-section, the design of the single-slot waveguide [9], the slant waveguide [10], the asymmetric rib waveguide [11, 12] and HPW [13] have been reported. Researchers from MIT [5], Waterloo University [10, 14, 15], Ghent University [16, 17] and IME [7, 18, 19] have done a lot of work on simulation and experiment, while researchers from Zhejiang University [11], Peking University [13] and other institutes [12, 20] also put forward some new design schemes (Fig. 4.1).

Figure 4.2 illustrates the schematic of various polarization-controlling devices. From the polarization diversity scheme, we can see that the combination effect of PBS and PR is to convert the input signal of any polarization state to two TE/TE outputs. Is there a device (we call it the polarization splitter-rotator, PSR) which can achieve above functions at once? In 2011, researchers from DTU first demonstrated a PSR by using the directional coupler in SOI waveguide with air cladding [21]. PSR doesn't mean simply cascading PBS and PR. It reduces the complexity of the design and lowers the difficulty of the process. We know that the processing of PR is more difficult and is not suitable for the CMOS process. In addition, it also greatly improves the insertion loss of the device, so it has a very broad prospect in the polarization diversity scheme.

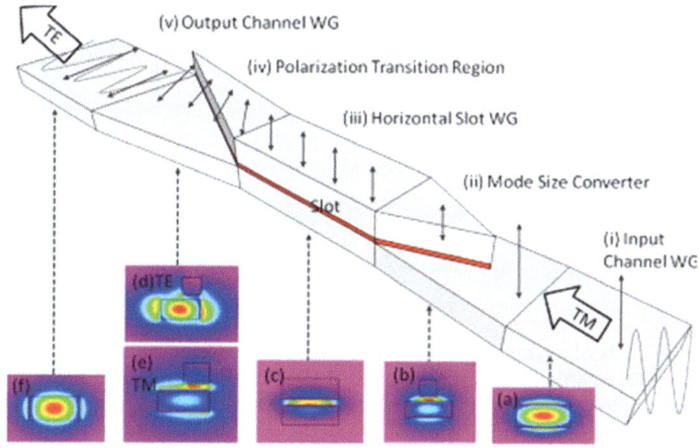

Fig. 4.1 A PR device in SOI platforms. Reprinted with permission from Ref. [7]. Copyright 2012, American Institute of Physics

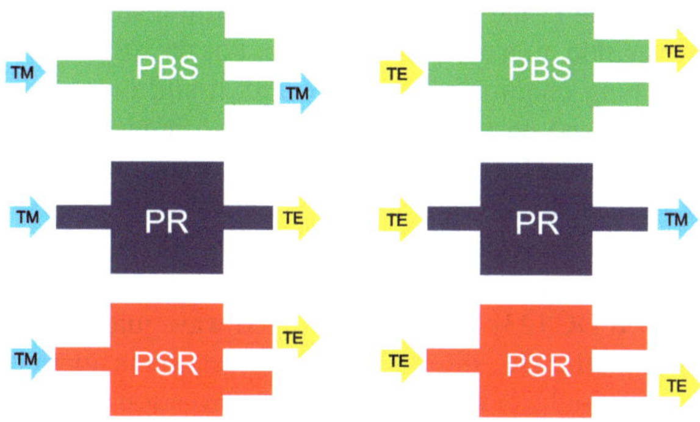

Fig. 4.2 Schematic of functions in different polarization-controlling devices

The mode conversion from TM to TE in PSRs origins from changing these two orthogonal modes into hybrid modes, which can be realized by asymmetric waveguides. For strip waveguides, the simplest way is to use an upper cladding of air. Because the material of the upper cladding (air) and the lower cladding (SiO_2) are not identical, the symmetry of the cross-section in the waveguide is broken. Considering the requirements of the CMOS process on the SiO_2 upper cladding, some new asymmetric waveguides need to be proposed, as shown in Fig. 4.3. These asymmetric waveguides have been partly used in CMOS-compatible Si-PSR. Here, we will review the recent progress of Si-PSR.

The early Si-PSR was mainly based on the strip waveguide with air upper cladding, which has the advantages of simple device structure and easy process. In 2011, Liu et al., from DTU reported the first DC-based Si-PSR in the world [21]. By optimizing the waveguide size, the effective refractive index of the quasi-TM mode

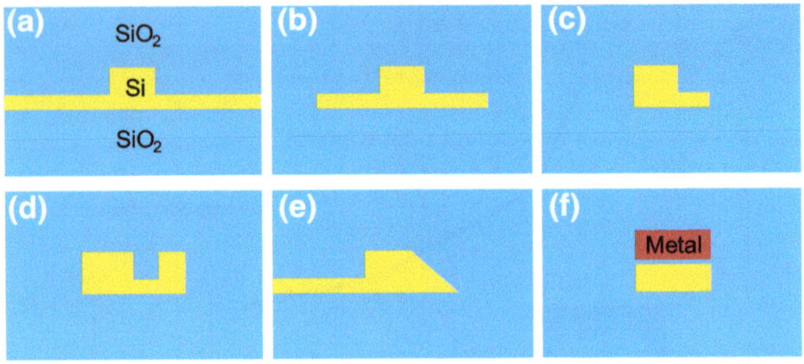

Fig. 4.3 Some kinds of asymmetric waveguide with SiO_2 upper cladding. **a** Rib waveguide. **b** Bi-level waveguide. **c** Partially-etched waveguide. **d** Slot waveguide. **e** slanted waveguide. **f** Hybrid plasmonic waveguide

in the wide waveguide is the same as that of the quasi-TE mode in the narrow waveguide, which satisfying the phase matching condition. Therefore, the quasi-TM mode in the wide waveguide will transfer to the quasi-TE mode in the narrow waveguide after a certain distance. For the quasi-TE mode input, because the phase-matching condition is not satisfied, it still outputs in the wide waveguide. Similar PSRs based on DC structures are also implemented by the above phase matching conditions. The length of the coupling region is only 36.8 μm, the insertion loss is 0.6 dB in the C band and the extinction ratio is 12 dB. In the same year, Ding et al., used the device to prove a 20 Gb/s NRZ-DPSK demodulator in the polarization diversity circuit [22]. In 2012, Ding et al., implemented a tolerant Si-PSR by using a taped DC [23]. They used a taped waveguide instead of a straight waveguide in a asymmetric DC, thus the mode conversion is aperiodic. Although the device length is increased to 140 μm, the mode conversion loss is less than 1 dB when the narrow waveguide width error is within 14 nm. In 2013, Ding et al., implemented a broadband and fabrication-tolerant Si-PSR device by using MMI couplers, phase shifters, mode converters, and Y-junctions [24]. First, the TM_0 mode is converted to TE_1 mode in a mode converter, and then the TE mode is divided into two channels by a Y-junction. By designing the appropriate phase, it can output in the desired port of the MMI coupler. The test results show that in the 100 nm wavelength range the insertion loss is less than 2.5 dB, the crosstalk is less than −12 dB, and the fabrication tolerance is larger than 50 nm. In the same year, they used the PSR device to prove a 40 Gb/s NRZ-DPSK demodulator in the polarization diversity circuit [25]. In 2011, Dai et al., used a mode converter and DC to propose a Si-PSR [26]. First, the TM_0 mode is converted to TE_1 mode by a three-section cascaded tapered mode converter, and then the TE_1 mode is coupled to the TE_0 mode in the narrow waveguide near the DC. The simulation results show that the device length is less than 100 μm, and the extinction ratio is better than 10 dB in 70 nm wavelength range. In 2012, Fei et al., proposed a Si-PSR by using a DC which consists of a slot waveguide and a strip waveguide [27]. The simulation results show that the coupling length is only 17.4 μm, and the extinction ratio is better than 15 dB in the wavelength range from 1515 to 1560 nm.

However, PSR devices without solid cladding are incompatible with most CMOS back-end process, which makes it difficult to integrate PSR devices with other SOI photonic devices. So the following researchers began to focus on the research of CMOS-compatible Si-PSR. In 2014, Sacher et al., at the University of Toronto reported a CMOS-compatible Si-PSR [28]. It has a bi-level tapered mode converter and an adiabatic coupler. The test results show that the device has crosstalk of less than −13 dB in 50 nm wavelength range. In addition, by using cascaded PBS devices as polarization filters, the crosstalk can be further reduced to −22 dB in 80 nm wavelength range. In 2014, Guan et al., from IME realized a Si-PSR with double etched DC [29]. The DC consists of a double etched waveguide and a strip waveguide. The phase matching condition is used to realize the mode conversion. The test results show that the device length is 27 μm, the mode conversion loss from TM to TE is less than 0.5 dB in the wavelength range from 1540 to 1570 nm, the TE insertion loss is less than 0.3 dB, and the crosstalk is less than −20 dB. In the same year, they also reported a Si-PSR consisting of a DC-PBS, a bi-level TM_0 to TE_1 mode converter, and a TE_1 to TE_0 MZI [30]. The PSR can operate at 1310 nm wavelength. The TM to TE mode conversion loss is about 2 dB in 40 nm wavelength range, and the crosstalk is less than −20 dB. In 2015, they also proposed a Si-PSR capable of operating at 1310 nm and 1550 nm at the same time, with 0.9 dB loss at 1310 nm wavelength and 1 dB loss at 1550 nm wavelength [31]. In 2014, Xiong et al., proposed a broadband Si-PSR scheme with high process tolerance by using double etched tapered DC [32]. The simulation results show that the TM to TE conversion loss is 0.09 dB in the C band, the crosstalk is −30 dB in the wavelength range above 160 nm, and the process tolerance is better than 50 nm.

In addition to these CMOS-compatible Si-PSRs, novel structures such as subwavelength grating waveguides or hybrid plasmonic waveguide can also be used in the PSR design. In 2014, Xiong et al., proposed a novel Si-PSR scheme to improve the fabrication tolerance by using subwavelength grating DC [33]. The simulation results show that the mode conversion loss from TM to TE is 0.13 dB at 1550 nm wavelength and is better than 0.4 dB in the C band. At the same time, the fabrication tolerance increases from ±3 nm to ±40 nm. In 2014, Yin et al., proposed an ultra-compact and low-loss Si-PSR by using Si-Ag HPW DC [34]. The simulation results show that the coupling length is 11.2 μm. The simulation results show that the loss is less than 0.42 dB in the 60 nm wavelength range.

Here we summarized the experimental results of Si PSRs in the Table 4.1.

In the following sections, we will take a look at a series of work we've done about the CMOS-compatible Si-PSR.

Table 4.1 Experimental summary of the previous Si-PSR (2011-2015.3)

Group	Time	Cladding	Structure[a]	Length (μm)	Performance[b]
Liu et al. [21]	2011	Air	Channel DC	36.8	0.6/–12/35
Ding et al. [23]	2012	Air	Tapered channel DC	100∼140	1/*/*
Ding et al. [24]	2013	Air	MC+YJ+PS+MMI	190	2.5/–12/100
Sacher et al. [28]	2014	SiO_2	Bi-level MC+Adiabatic coupler	400	1.5/–13/50
Guan et al. [29]	2014	SiO_2	Double-etched DC	27	0.5/–20/30
Guan et al. [30]	2014	SiO_2	Channel DC+Bi-level MC+MZI	56	2/–20/40

[a]MC: mode converter, YJ: Y-junction
[b]IL (dB) /XT (dB) /Bandwidth (nm)
*denotes no performance characterization from the reference in the table

4.2 Compact Polarization Splitter-Rotator Based on an Rib Asymmetric Directional Coupler

In this section, we propose a Si-PSR design based on rib directional coupler with the upper cladding SiO_2. Rib waveguide is asymmetric in its horizontal direction of the cross-section, so the mode hybridization can be achieved in the case of SiO_2 upper cladding. This is much more stable than the device without upper cladding, and it can also greatly improve the application scenario of the polarization diversity mechanism. Moreover, the coupling between the rib waveguides is stronger than that between the strip waveguides, which is beneficial to the realization of the ultra-compact PSR device.

4.2.1 Single Si-PSR Based on Rib Asymmetric Directional Coupler

The structure of Si-PSR based on rib ADC is shown in Fig. 4.4. Its main part is a ADC. The width of two waveguides are W1 and W2 respectively, and the waveguide gap is G and the length is Lc. In the output section, an S-bend waveguide with a length of Ls is used to remove the coupling between the two ports. The operation principle of Si-PSR based rib ADC is similar to other PSR devices based on ADC. In order to achieve efficient coupling between two orthogonal modes TE_0 and TM_0, the phase matching condition is required. In other words, the effective refractive index of the TM_0 mode of the input waveguide needs to be the same as that of the TE_0 mode of the adjacent waveguide. Therefore, at a certain coupling length Lc, the input TM_0 mode will be completely converted to the TE_0 of the adjacent waveguide and output at the cross port. For the input TE_0 mode, because there is no such phase matching condition, it will propagate along the input waveguide and output at the through port.

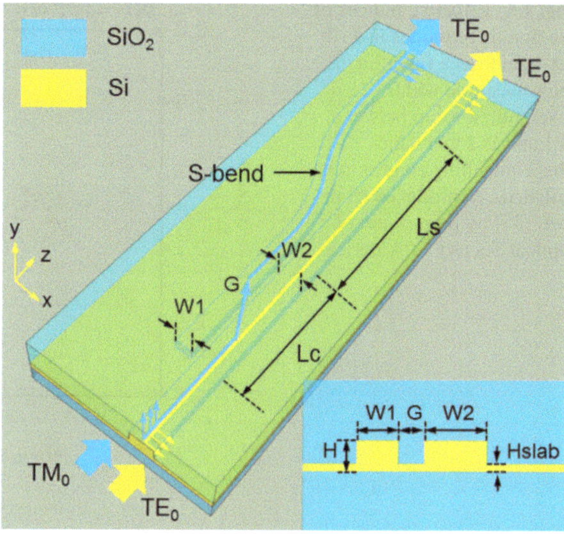

Fig. 4.4 Schematic of the proposed PSR based on a rib DC. The cross-section of the DC is also presented. Reproduced from Ref. [35] by permission of Optical Society of America

In order to find the waveguide parameters W1 and W2 that can meet the phase matching conditions, we used the two-dimensional simulation software FIMMWAVE to calculate the effective refractive index of the first three guide modes in a SOI rib waveguide. The top silicon thickness (H) of the rib waveguide we chose is 250 nm, and the slab thickness (Hslab) is 50 nm. The refractive indices of Si and SiO_2 materials are 3.455 and 1.445, respectively. The effective refractive index obtained at 1550 nm wavelength varies with the width of the waveguide, as shown in Fig. 4.5. We find that the waveguide parameters satisfying the phase matching conditions can be obtained in a large range of W1 and W2. Here, we chose W1 = 262 nm and W2 = 500 nm to ensure that the effective refractive index of the TM_0 mode in the wide waveguide is the same as that of the narrow waveguide, so an efficient mode conversion will be found between the two modes. For the TE_0 mode in a wide waveguide, it is difficult to be affected by the narrow waveguide, because its effective refractive index is very different from that of any guided mode in the narrow waveguide. Therefore, for the TE_0 mode in the wide waveguide, it no longer satisfies the phase matching condition.

We also extract the relationship between waveguide widths W1 and W2 when the phase matching condition is satisfied. W1 increases linearly with W2, and the slope is about 1/3. This also implies that the phase matching condition is more sensitive to the change of W1. Nevertheless, our proposed PSR has a better fabrication tolerance than the PSR based on air-cladding ADC [21]. The slope of the latter curve is 1/12. In other words, for the PSR with air-cladding, 1 nm width variation of the narrow waveguide requires about 12 nm adjustment of the width of the wide waveguide in order to satisfy the phase-matching condition,

Here, we first define some important parameters to describe the PSR performance, i.e., insertion loss (IL), extinction ratio (ER) and crosstalk (XT). These three parameters can be defined in terms of input polarization state or output port. To facilitate

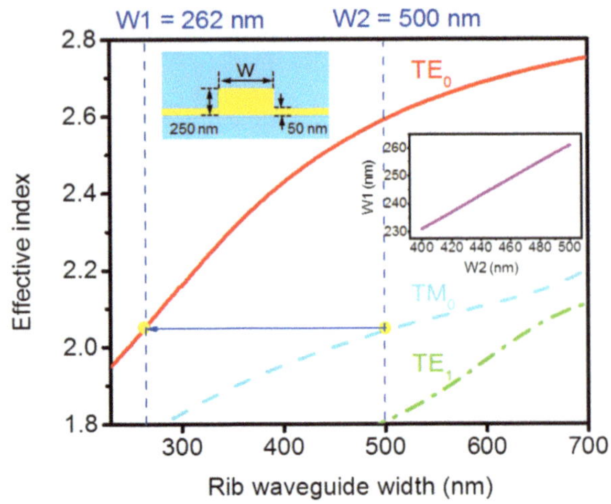

Fig. 4.5 Effective indices of the first three modes in a rib waveguide as a function of the waveguide width. Inset: relationship between W1 and W2 when the phase-matching condition is satisfied. Reproduced from Ref. [35] by permission of Optical Society of America

the description, we assume that for the input TE mode, it will output at port 1. For input TM mode, it will output at port 2 as the TE mode. P_{Mode}^{Port} is defined as the power of MODE at the PORT. Here, we define these parameters according to the input polarization state.

$$IL_{TE} = -10\log_{10}(P_{TE_0}^{Port1}/P_{TE_0}^{PortIN}) \quad (4.1)$$

$$IL_{TM} = -10\log_{10}(P_{TE_0}^{Port2}/P_{TM_0}^{PortIN}) \quad (4.2)$$

$$ER_{TE} = 10\log_{10}(P_{TM_0}^{Port1}/P_{TE_0}^{Port1}) \quad (4.3)$$

$$ER_{TM} = 10\log_{10}(P_{TM_0}^{Port2}/P_{TE_0}^{Port2}) \quad (4.4)$$

$$XT_{TE} = 10\log_{10}(\max(P_{TE_0}^{Port2}, P_{TM_0}^{Port2})/P_{TE_0}^{Port1}) \quad (4.5)$$

$$XT_{TM} = 10\log_{10}(\max(P_{TE_0}^{Port1}, P_{TM_0}^{Port1})/P_{TE_0}^{Port2}) \quad (4.6)$$

In order to optimize this PSR device, we used the three-dimensional simulation software FIMMPROP based on eigenmode expansion (EME) to calculate the mode conversion efficiency of the whole device. In the simulation, we choose the wavelength of 1550 nm and the gap between two waveguides of 150 nm, and the output part slowly increases to 1 μm. The S-bend waveguide length Ls is chosen as 15 μm. Figure 4.6a, b show the relationship between the mode conversion efficiency and the coupling length Lc in the cross-port and through-port, respectively. We find that when the coupling length is about 9 μm, the mode conversion efficiency from the input TM_0 mode to the TE_0 mode reaches a maximum of 97%, and the corresponding extinction ratio (ER) are 20 dB and 26.7 dB, respectively. Moreover, the extinction ratio of the cross-port is not sensitive to the change of the coupling length Lc, mainly

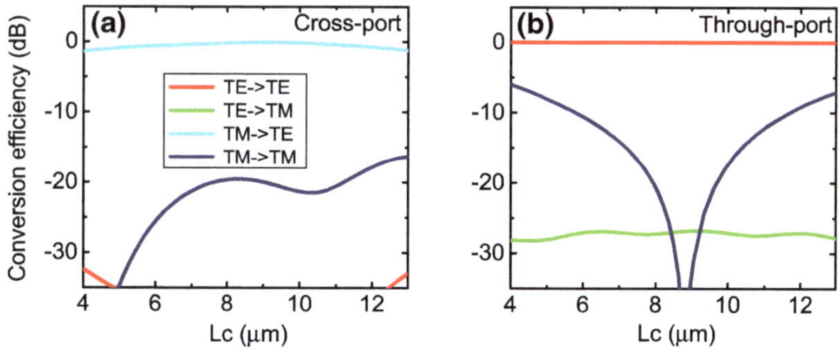

Fig. 4.6 a–b Conversion efficiency as a function of coupling length (Lc) in cross-port and through-port, respectively. The conversion efficiency below 35 dB is not shown. Reproduced from Ref. [35] by permission of Optical Society of America

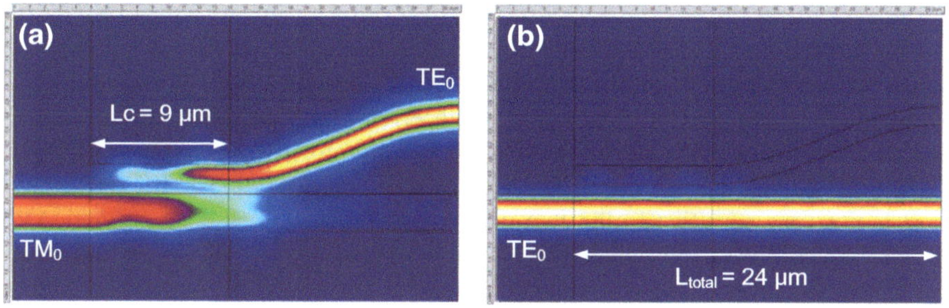

Fig. 4.7 a–b Mode propagation when the input is (**a**) the TM_0 mode and (**b**) the TE_0 mode at 1550 nm wavelength. Reproduced from Ref. [35] by permission of Optical Society of America

because the narrow waveguide does not support the transmission of TM_0 guided mode, so that it is equivalent to a mode filter that will filter the converted TM_0 mode.

Figure 4.7a, b show the mode propagation in the whole device when the inputs are the TM_0 and TE_0 mode, respectively. As you can see, for the input TM_0 mode, most of the power has been transferred to the TE_0 mode of the cross-port. The input TE_0 mode propagates directly through the input waveguide, while there is almost no coupling between the adjacent waveguides.

Figure 4.8a, b show the wavelength dependence of the device performance in the cross port and through port, respectively. For the cross port, when the wavelength is between 1500 and 1600 nm, the IL is from 0.14 to 1.75 dB, and the ER is better than 19 dB. However, because the effective refractive index of the mode is wavelength dependent, the phase matching condition is no longer satisfied when the operating wavelength is changed, and the mode conversion efficiency from TM_0 to TE_0 is reduced. These residual TM_0 mode will continue to transmit in the input waveguide, resulting in a decrease in the ER of the through port. Nevertheless, for a straight through port, the ER in the C band is also better than 9 dB.

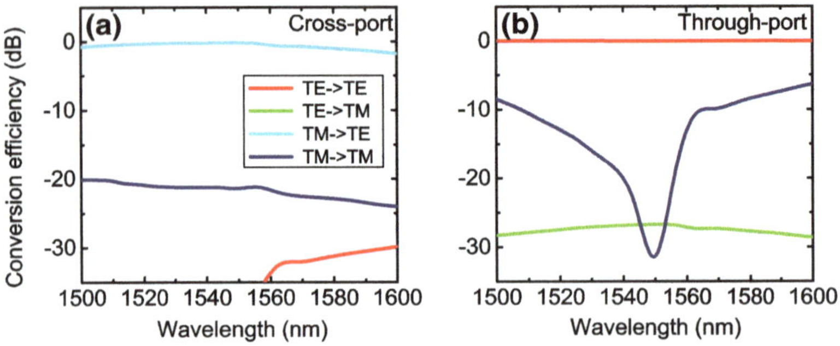

Fig. 4.8 **a–b** Conversion efficiency as a function of wavelength in cross-port and through-port, respectively. The conversion efficiency below 35 dB is not shown. Reproduced from Ref. [35] by permission of Optical Society of America

We further analyzed the effect of fabrication deviations on device performance when the wavelength is 1550 nm. Figure 4.9a–d show the effect of waveguide gap deviation ΔG and narrow waveguide width error ΔW1 on device performances. Because the width of the narrow waveguide is more sensitive to the phase matching condition, we assume that the width of the waveguide is fixed. The ER of the cross port is not very sensitive to the two deviations. When ΔG = ±50 nm and ΔW1 = ±15 nm, ER with more than 12 dB can be obtained, showing larger fabrication tolerance. However, under the same condition, the performance in the through port will decrease significantly. Nevertheless, when the waveguide gap deviation is ±50 nm, the ER is still better than 10 dB. Compared with strip waveguides, there is a new degree of freedom Hslab in the vertical direction of rib waveguide, which has great influence on mode properties. Figure 4.9e, f show the effect of ΔHslab on device performance respectively. This error has great influence on the performance of the device, so the etching process is very important here. Nevertheless, when ΔHslab is in the range from 55 to 50 nm, the performance is acceptable. Another fabrication error occurs in the oxidation deposition of the upper cladding, which may lead to changes in the refractive index of the upper cladding. Figure 4.9g, h show the effect of this error on performances. We find that when the refractive index error is ±5%, ER is higher than 10 dB.

For the silicon photonic devices based on rib waveguide, a very important problem to consider is the mode conversion between the waveguide and the strip waveguide. In silicon photonic integration, the strip waveguide is more common than the rib waveguide because the strip waveguide has a stronger restriction on light, which is beneficial to the realization of the curved waveguide with small radius and the enhancement of the density of the photon integration. Figure 4.10 shows a double-etched tapered mode converter with a length of L_{tp} for rib waveguide and strip waveguide. On one hand, the tapered mode converter needs to be long enough to ensure adiabatic mode conversion. On the other hand, since the waveguide symmetry has been broken at this time, there may be a mode conversion between the TM_0

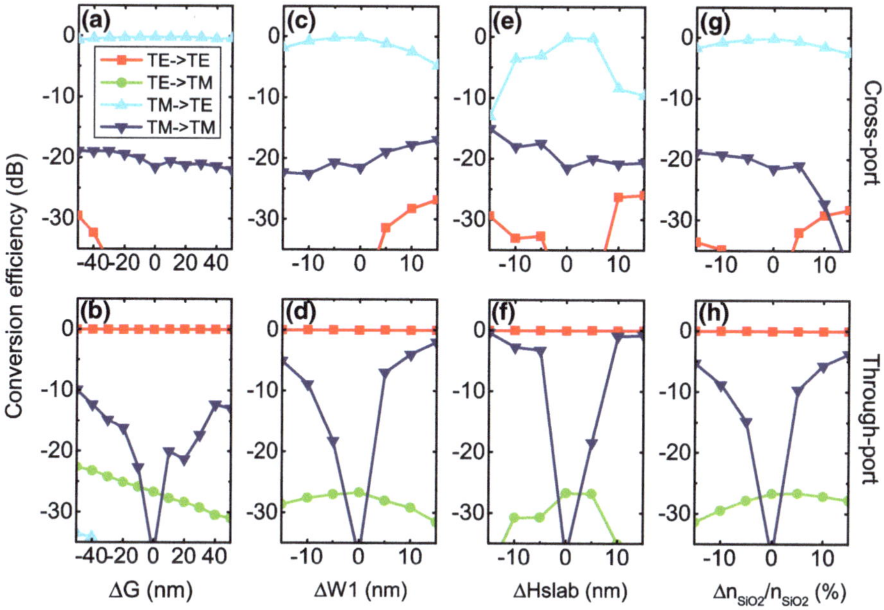

Fig. 4.9 Fabrication tolerance analysis for the (**a**)–(**b**) the gap variation ΔG, (**c**)–(**d**) width variation ΔW1 of the narrow waveguide, (**e**)–(**f**) slab height variation ΔHslab and (**g**)–(**h**) variation Δn_{SiO2} of the refractive index of the upper-cladding. The conversion efficiency below 35 dB is not shown. The wavelength is 1550 nm. Reproduced from Ref. [35] by permission of Optical Society of America

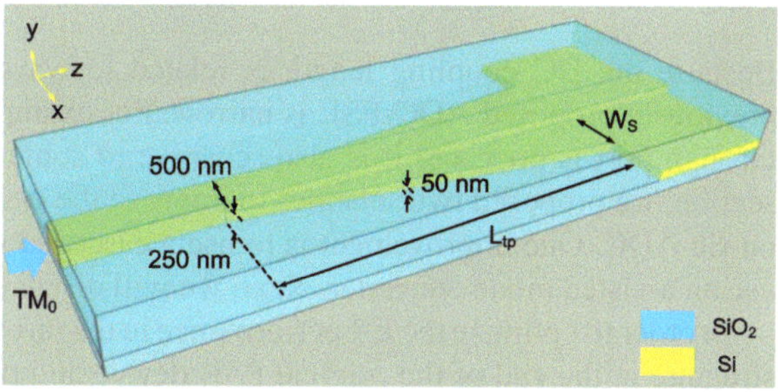

Fig. 4.10 Schematic of the taper between the strip waveguide and the rib waveguide. Reproduced from Ref. [35] by permission of Optical Society of America

mode and the TE_0 mode, that is to say, the tapered waveguide may change the input polarization, thus limiting the application of the PSR [36].

We calculated the relationship between the transmission of the input TM_0 and the width Ws at the end of the tapered waveguide when the taper length is L_{tp} = 10 μm and the wavelength is 1550 nm, as shown in Fig. 4.11a. It can be found that the transmission increases with Ws. When Ws increases further, because L_{tp} is not long enough at this time, the transmission will decrease slightly. But there is no polarization transition. We chose Ws = 0.3 μm and L_{tp} = 10 μm and calculated the wavelength dependence of the TM_0 transmission, as shown in Fig. 4.11a. It is found that transmission is close to 1 in the wavelength range from 1500 to 1550 nm.

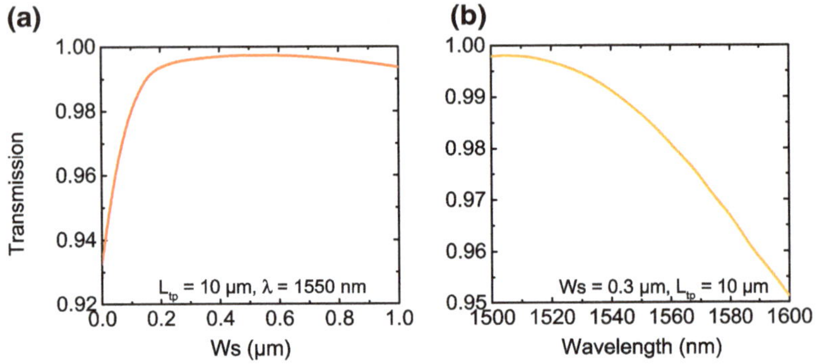

Fig. 4.11 **a** The transmission as a function of Ws when $L_{tp} = 10\ \mu m$. **b** The transmission as a function of wavelength when Ws = 0.3 μm and $L_{tp} = 10\ \mu m$. Reproduced from Ref. [35] by permission of Optical Society of America

When the wavelength exceeds 1550 nm, the transmission will gradually decrease. Nevertheless, transmission is greater than 95% in the whole 100 nm wavelength range.

4.2.2 Improved Si-PSR Based on Rib ADC by Using Cascaded Structure or Mode Convener

Because the DC coupling length is related to wavelength, the bandwidth of the previously proposed ADC-PSR is narrow. For example, the ER performance in the through port is poor and it is also sensitive to some fabrication deviations. In this section, we proposed two schemes to improve the performance of the Si-PSR based on rib ADC. One improvement is to use a cascaded structure and another one is to use an assisted mode converter. Next, we will discuss about it in detail.

In order to optimize the ER performance in the through port, we cascaded a similar structure at the end of the original PSR device and optimized the performance at a wavelength of 1600 nm. The schematic of the cascaded PSR is shown in Fig. 4.12. The final design parameters of the whole device are as follows: W1 = 262 nm, W2 = 500 nm, W3 = 250 nm, G = 150 nm, Gout = 1 μm, Lc1 = 11 μm, Lc2 = 7 μm, and Ls1 = Ls2 = 15 μm.

We calculated the performance of the whole device in the cross-port, as shown in Fig. 4.13. It can be found that at the wavelength of 1600 nm, ER has increased from 6.3 dB at the original ADC to 15.7 dB.

We have calculated the mode propagation at this wavelength, as shown in Fig. 4.14. You can see that the unconverted TM_0 mode in the first ADC is coupled to the adjacent waveguide in the second ADC, which proves the improvement of ER performance.

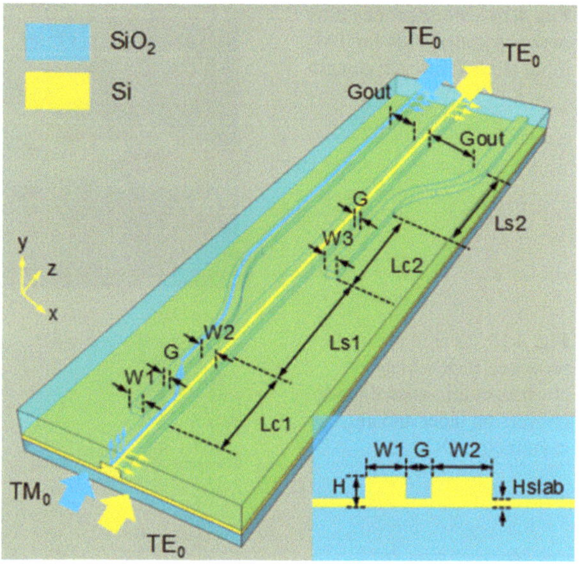

Fig. 4.12 Schematic of the cascaded PSR, which has two DCs with optimal coupling length at 1550 nm and 1600 nm wavelengths, respectively

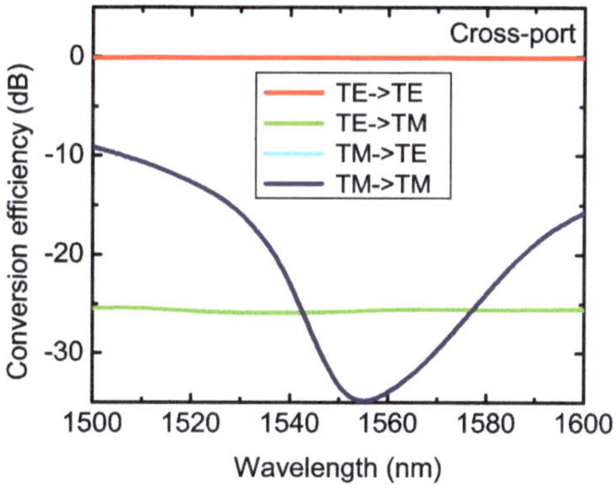

Fig. 4.13 Improved conversion efficiency as a function of wavelength in the cross-port of the cascaded PSRs

Compared to a single Si-PSR based on rib ADC, such a cascaded PSR provides more design freedom, and the bandwidth and other performances can be designed more conveniently.

Another optimization method is to add an assisted mode converter before the ADC. Figure 4.15 shows the improved PSR based on rib waveguide. It consists of a three-order linear tapered mode converter and a ADC. The widths of the three-order linear tapered mode converter are W_0, W_1, W_2 and W_3, respectively. The length in each section is L_{tp1}, L_{tp2} and L_{tp3}, respectively. The ADC consists of two waveguides with a width of W_3 and W_4 respectively and a gap of G. In the output section, the gap between the two waveguides will increase gradually to G_{out} in the cross port and through port, and the waveguide width is W_0. The device is also designed on the SOI

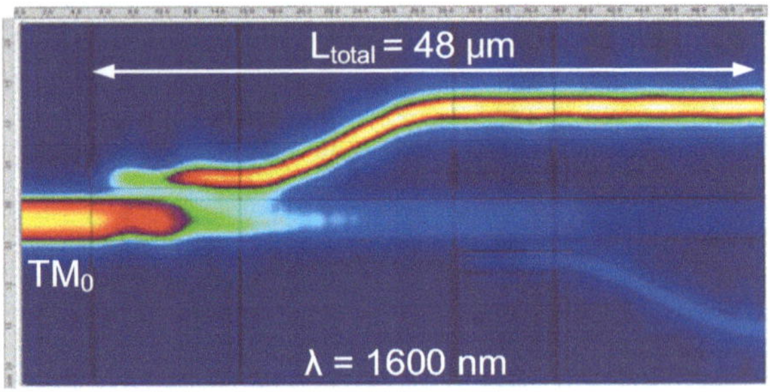

Fig. 4.14 Mode propagation in the cascaded PSR for TM_0 input at 1600 nm wavelength

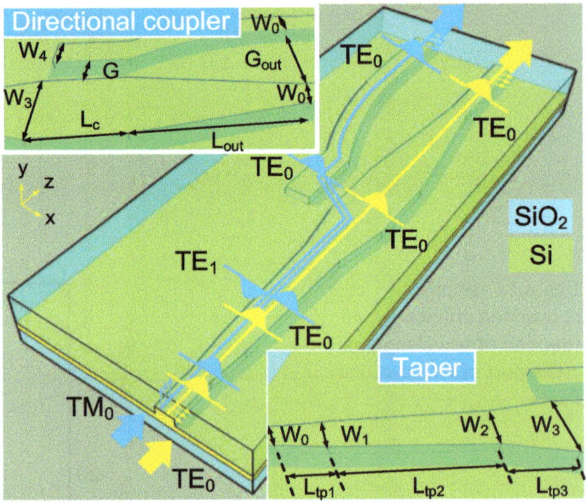

Fig. 4.15 Schematic of the proposed PSR which consists of an assisted mode conversion taper and an asymmetric DC

wafer with H = 220 nm top silicon layer, H_{slab} = 50 nm slab waveguide and SiO_2 upper cladding. The working principle of this improved PSR device is described as follows:

- On one hand, for the input TM_0 mode, it will evolve gradually into the TE_1 mode in the tapered mode converter. Because the cross-section symmetry of the waveguide is broken in the rib mode converter, there is a hybrid mode in some places. Next, the TE_1 mode propagates into the ADC region. As long as the effective refractive index of the TE_1 mode is the same as that of the TE_0 mode in the narrow waveguide, it will be coupled to the TE_0 mode in the adjacent narrow waveguide.
- On the other hand, for the input TE_0 mode, it will not be converted to any other guided mode in the tapered mode converter, and then propagates through the ADC region. There is almost no coupling in the adjacent waveguide. This is mainly because the phase matching condition is not satisfied at this time.

In order to determine the width of each waveguide, we first used a two-dimensional simulation software FIMMWAVE to calculate the relationship between the effective refractive index of the first three guided modes and the width of the waveguide

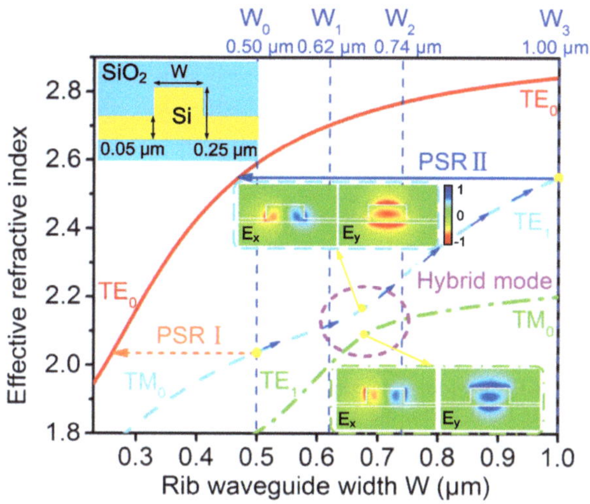

Fig. 4.16 Effective refractive indices of the first three modes in the waveguide cross-section. Insets: the profiles of the first- and second-order modes in the hybrid mode region

when the wavelength is 1550 nm, as shown in Fig. 4.16. The refractive index of the material in the simulation is based on the wavelength-dependent model provided by the software. It can be found that when the waveguide width varies from $W_1 = 0.62$ μm to $W_2 = 0.74$ μm, mode hybridization occurs in the second and third modes. The field distribution of the two modes in the mode hybridization region shows that the intensity of the secondary component (E_x or E_y) can be comparable to the main component (E_y or E_x) at this time, which may lead to the conversion between the two modes. Therefore, in order to convert the input TM_0 into TE_1 in the tapered mode converter and avoid the mode hybridization at the end, the three sections of the waveguides can be changed from $W_0 = 0.50$ μm to $W_1 = 0.62$ μm, from $W_1 = 0.62$ μm to $W_2 = 0.74$ μm, and from $W_2 = 0.74$ μm to $W_3 = 1$ μm, respectively. In the PSR (called PSR-I) that we have not improved before, the width of the ADC is 0.50 μm and 0.262 μm, respectively. The TM_0 mode in the wide waveguide can be converted to the TE_0 mode in the narrow waveguide. However, when the width of the waveguide is close to 0.262 μm, the effective refractive index curve of TE_0 has a large slope. Therefore, a slight change in the waveguide size or working wavelength leads to a large change in the effective refractive index, which leads to a significant reduction in the performance of the device. Therefore, we choose a wider width of $W_3 = 1$ μm and corresponding $W_4 = 0.469$ μm for the improved ADC. In this way, a small slope of the TE_0 curve can be obtained, thus potentially improving the device performance and fabrication tolerance. Larger W_3 will achieve a smaller curve slope, but will increase the length of the mode converter.

We used a three dimensional simulation software FIMMPROP based on the eigenmode expansion (EME) algorithm to optimize the mode convention efficiency. Figure 4.17a shows the relationship between the conversion efficiency of the input TM_0 to other modes and L_{tp2} when L_{tp1} changes from 1 to 10 μm, Here, $L_{tp3} = L_{tp1}(W_3 - W_2) / (W_1 - W_0)$. It can be found that for a given L_{tp1}, the conversion effi-

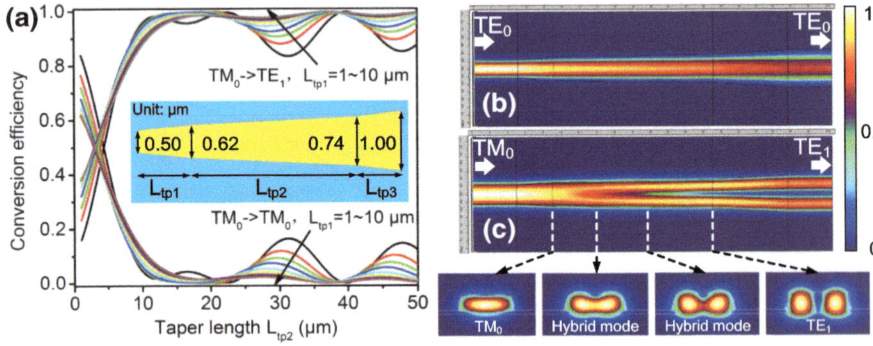

Fig. 4.17 **a** Mode conversion efficiency for the input TM_0 mode as a function of L_{tp2} with L_{tp1} varying from 1 μm (black) to 10 μm (dark yellow). **b–c** Mode propagation for the input TE_0 and TM_0 mode in the taper

ciency of almost 1 of TM_0 to TE_1 can be obtained as long as an appropriate L_{tp2} is selected. Moreover, a larger L_{tp1} reduces the ripple on the conversion curve. We consider the efficient mode conversion and the minimum device length. The tapered mode converter chooses the design parameters as follows: $L_{tp1} = 4$ μm, $L_{tp2} = 18$ μm. Figure 4.17b, c show the mode propagation in a tapered mode converter after the optimization for the TE_0 and TM_0 inputs, respectively. It can be found that the input TM_0 mode will slowly evolve into a hybrid mode and eventually output in TE_1 mode.

Next we will optimize the mode conversion efficiency in the ADC. Figure 4.18a shows the relationship between the mode conversion efficiency of the input TE_1 to the TE_0 in the cross port and the TM_0 in the through port with coupling length L_c and S bend length L_{out}. If the phase matching condition is satisfied, the input TE_1 will be mainly coupled into the TE_0 mode in the next waveguide. However, with the decrease of the waveguide width from W_3 to W_0, the remaining TE_1 power will be converted to TM_0 again. When the coupling length L_c is about 18.2 μm, the maximum mode conversion efficiency can be obtained. Moreover, the largest mode conversion efficiency will increase with the increase of L_{out}, mainly because the mode transition loss of S-bend waveguide is reduced. To ensure high conversion efficiency and proper device length, we choose $L_{out} = 6$ μm. Figure 4.18b, c show the mode propagation in DC in the case of input TE_0 and TE_1, respectively. It is clear to see that the mode efficiency between the input TE_1 and the TE_0 in the narrow waveguide is high, while the input TE_0 passes through ADC directly, and is basically not coupled with the adjacent waveguide.

Figure 4.19 shows the propagation of different input modes in the whole device when the wavelength is 1550 nm. The simulation results confirm that our design for each part is reliable. The total length of the device is 54.9 μm, which is slightly larger than that reported previously. A potential optimization is to adopt tapered mode con-

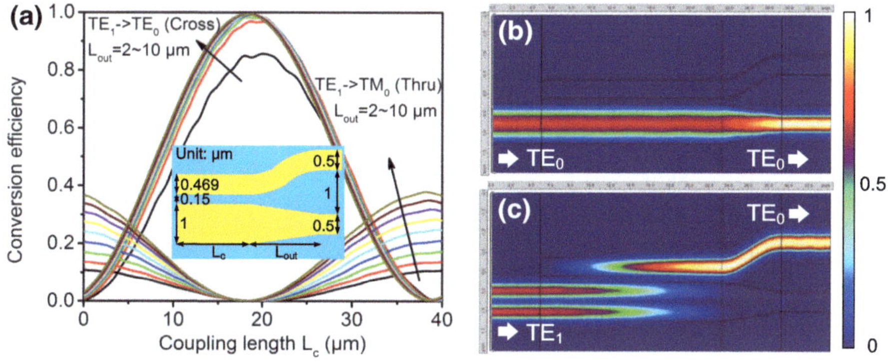

Fig. 4.18 **a** Mode conversion efficiency from TE_1 to TE_0/TM_0 in the cross/through port in the ADC as a function of L_c with L_{out} varying from 2 μm (black) to 10 μm (dark yellow). **b–c** Mode propagation in the ADC for different incoming polarizations

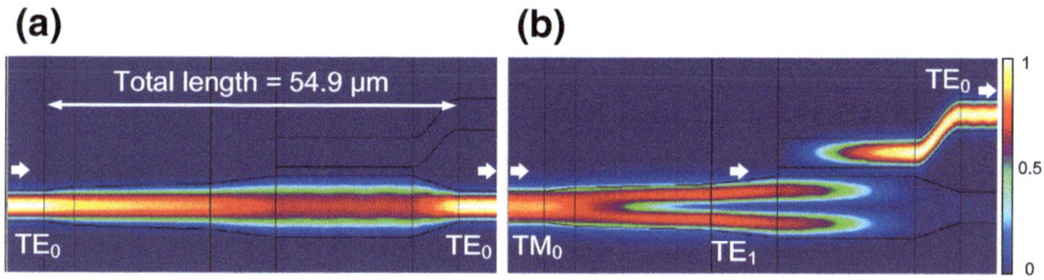

Fig. 4.19 Mode propagation in the improved PSR device at 1550 nm wavelength when the inputs are (**a**) the TE_0 and (**b**) TM_0 mode, respectively

verters with other shapes, such as sinusoidal or parabolic [37], or multi-variable optimization techniques such as genetic algorithms or particle swarm optimization [38].

The performance of PSR devices can be described by IL or XT, specifically referring to the contents of the previous section. The definition of XT takes into account the fact that the residual TE_1 in the output region will be re converted to TM_0 as the waveguide width decreases to W_0. Figure 4.20a, b show the wavelength dependence of the performance of modified ADC-PSR (called PSR-II) for TE_0 and TM_0 inputs respectively. We compare the performance of the previous ADC-PSR in the figure. It can be found that when the input is TE_0, the performance of the two devices is very stable in a large wavelength range. In this case, the TE_0 mode from the input port directly passes through the PSR device and almost does not interact with the adjacent waveguide. However, in the case of the input TM_0, PSR-II performs significantly better in terms of IL and XT. This is mainly due to the wider DC waveguide. PSR-II has IL less than 1 dB and XT less than −10 dB in the S, C and L optical communication band.

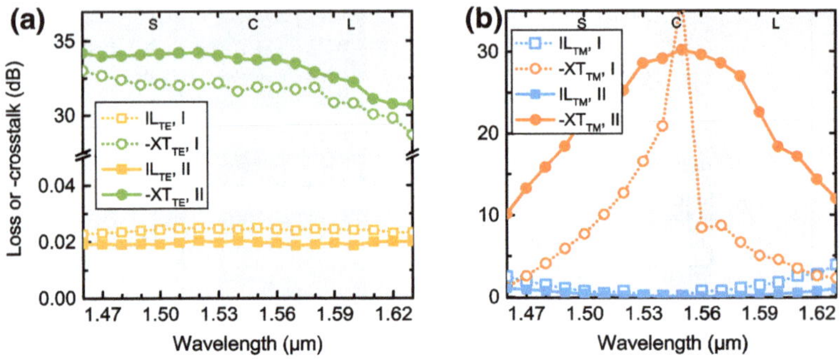

Fig. 4.20 Wavelength dependence of the improved PSR-II performances for the input TE_0 and TM_0 mode, respectively. The loss and crosstalk of the normal PSR-I are also shown here as a comparison

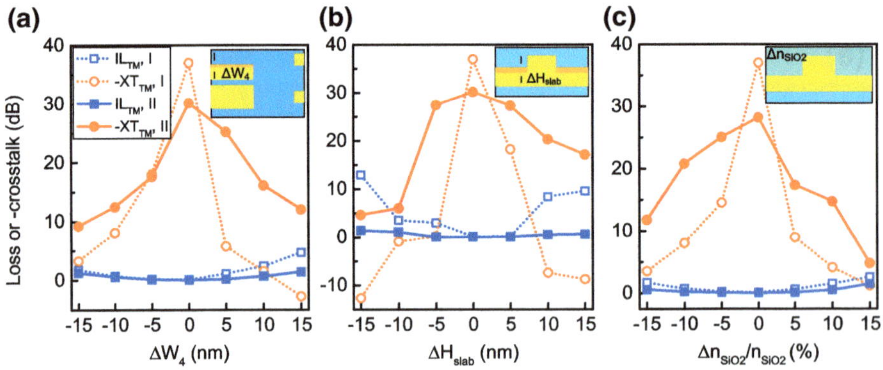

Fig. 4.21 Fabrication tolerance analysis to the deviations in (**a**) narrow waveguide width ΔW_4 of the DC, (**b**) ΔH_{slab}, and (**c**) refractive index of the upper-cladding Δn_{SiO2}, respectively

We also compared the sensitivity of these two devices to the ADC's narrow waveguide width error ΔW_4, the slab height error ΔH_{slab} and the refractive index error Δn_{SiO2} of the upper cladding material, as shown in Fig. 4.21a–c, respectively. Because the device with TE_0 input is very stable, we only show the device performance under TM_0 input. It can be found that these errors have a great impact on the performance of PSR-I devices. For ΔW_4 and ΔH_{slab}, even small errors of 5 nm will make the device fail to work properly. In the improved device PSR-II, the tolerance of the process has been improved. We believe that by further optimizing the process, such devices can exhibit good performance in practice.

4.3 Ultra-Broadband Si-PSR Based on Mode-Evolution Tapers and an Asymmetric Y-Junction

In Chap. 3, we proposed a scheme to achieve ultra-broadband PBS by using the mode-sorting function of cascaded asymmetric Y-junctions. Here, we also use the asymmetric Y-junction and the assisted mode converter to realize the function of PSR. The PSR device based on ADC requires strict phase matching and large device size, and the mode conversion in PSR based on asymmetric Y-junction and mode converter is all based on mode evolution instead of mode interference, so it has the characteristics of large bandwidth, high process tolerance and so on.

Figure 4.22 shows the schematic of the Si-PSR based on the three-order mode converter and the asymmetric Y-junction. The device is also designed on the SOI rib waveguide with 250 nm top silicon, 50 nm slab height, and SiO_2 upper cladding. The widths of the three-order linear mode converter on each end are W_0, W_1, W_2 and W_3, respectively. The length of each order are L_{tp1}, L_{tp2} and L_{tp3}, respectively. The width of the root waveguide of the asymmetric Y-junction is the same as that of the end of the tapered mode converter, and the widths of the two waveguide arms are W_w and W_n respectively, which satisfies the relationship of $W_w + W_n = W_3$. The length of the Y-junction is L_y, and the gap between the two ports at the end is G_{out}. The working principle of the whole device is as follows:

- For the input TM_0 mode, it is first converted into TE_1 mode in the three-order tapered mode converter, because the waveguide symmetry is broken in the rib mode converter. And there is a hybrid mode in some places. Then, by using the mode-sorting function of the asymmetric Y-junction, the TE_1 mode can be evolved into the TE_0 mode of the narrow arm waveguide and output in port 2.
- For the input TE_0 mode, it will not be converted into any other modes in the mode converter, but directly go through the mode converter and evolve into the TE_0 mode in the wide arm of the Y-junction, and finally output in port 1.

In order to determine the width of each waveguide, we first used a two-dimensional simulation software FIMMWAVE to calculate the relationship between the effective refractive index of the first three guided modes and the waveguide width when the wavelength is 1550 nm, as shown in Fig. 4.23. In the simulation, the refractive index of Si is 3.476, and the refractive index of SiO_2 is 1.528. It can be found that when the waveguide width varies from $W_1 = 0.62$ μm to $W_2 = 0.74$ μm, the mode hybridization occurs in second and third modes. Therefore, in order to convert the input TM_0 into TE_1 in the tapered mode converter and avoid the mode hybridization at the end, the widths of the three sections of the waveguides can be changed from $W_0 = 0.50$ μm to $W_1 = 0.62$ μm, from $W_1 = 0.62$ μm to $W_2 =$ 0.74 μm, and from $W_2 = 0.74$ μm to $W_3 = 1.00$ μm. In addition, according to the mode-sorting effect of the asymmetric Y-junction, it is found that the width W_n of the asymmetric Y-junction in the narrow arm can be obtained within the range of 270 nm to 400 nm. The minimum of W_n is decided by considering that $|Neff_{W3,TE1}\text{-}Neff_{Ww,TM0}| < |Neff_{W3,TE1}\text{-}Neff_{Wn,TE0}|$, when $W_n < 270$ nm, $W_w > 530$ nm. Here,

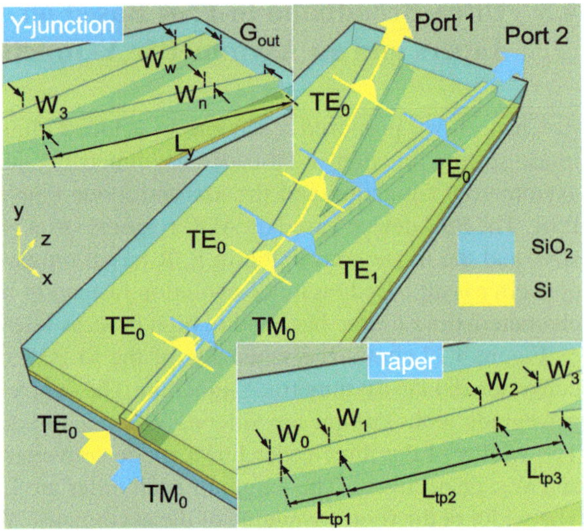

Fig. 4.22 Schematic of the proposed PSR consisting of mode-evolution tapers and a mode-sorting asymmetric Y-junction. Reproduced from Ref. [39] by permission of Optical Society of America

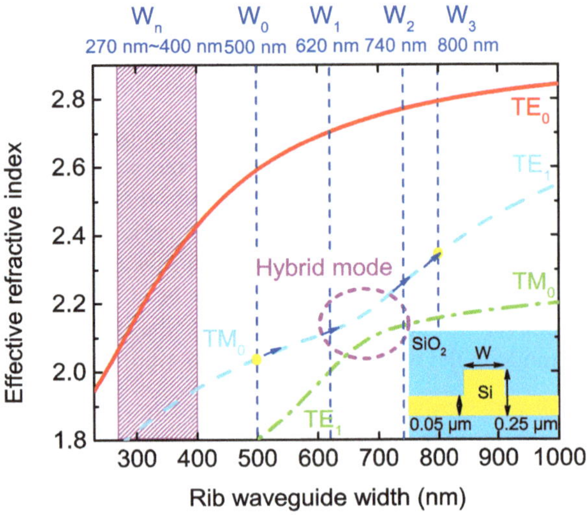

Fig. 4.23 Effective indices of the first three modes in a rib waveguide as a function of the waveguide width. Reproduced from Ref. [39] by permission of Optical Society of America

$Neff_{w,mode}$ is the effective refractive index of MODE when the waveguide width is W. The expression means that the TE_1 mode input from the root waveguide will evolve into the TM_0 mode in the wide arm instead of the TE_0 mode in the narrow arm, resulting in an unwanted mode-sorting.

We used a three dimensional simulation software FIMMPROP based on the eigenmode expansion (EME) algorithm to optimize the mode conversion efficiency. Figure 4.24a shows the relationship between the conversion efficiency of the input TM_0 to other modes and L_{tp2} when L_{tp1} changes from 1 to 10 μm. Here, $L_{tp3} = L_{tp1}(W_3\text{-}W_2) / (W_1\text{-}W_0)$. It can be found that for a given L_{tp1}, the conversion effi-

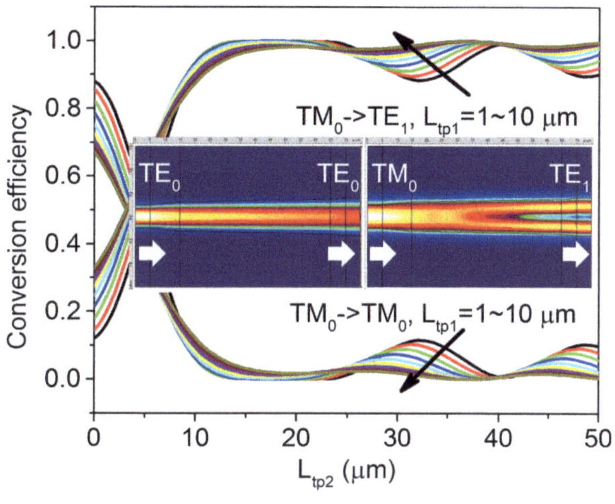

Fig. 4.24 Mode conversion efficiency for the input TM_0 mode as a function of L_{tp2} for L_{tp1} varying from 1 μm (black) to 10 μm (dark yellow). Reproduced from Ref. [39] by permission of Optical Society of America

ciency of TM_0 to TE_1 is close to 1 as long as an appropriate L_{tp2} is selected. Moreover, a larger L_{tp1} reduces the ripple on the conversion curve, which is beneficial to improve the fabrication tolerance. In order to achieve an efficient mode conversion in a large wavelength range, the design parameters of the tapered mode converter is chosen as follows: $L_{tp1} = 10$ μm, $L_{tp2} = 50$ μm and $L_{tp3} = 5$ μm. The internal diagram of Fig. 4.24 shows the mode propagation in the tapered mode converter after the optimization for TE_0 and TM_0 inputs, respectively. It can be found that the input TM_0 mode will slowly evolve into a hybrid mode and eventually output as TE_1 mode.

Figure 4.25 shows the relationship between the mode conversion efficiency and the Y-junction length L_y with different width of narrow arm when the wavelength is 1550 nm. Here, we consider the TE_0 mode and the TE_1 mode converted from the TM_0 mode. In the simulation, the gap G_{out} in the output port is 1 μm. When the width of the narrow arm changes between 284 and 374 nm, the exact mode sorting and high conversion efficiency can be obtained for different input modes, as shown in Fig. 4.25a, b. Moreover, the conversion efficiency will increase with the increase of L_y. However, when $W_n = 274$ nm, there will be some obvious ripples in the mode conversion efficiency curve, as shown in Fig. 4.25c. This is because at this time $|Neff_{W3,TE1}\text{-}Neff_{Ww,TM0}|$ is very close to $|Neff_{W3,TE1}\text{-}Neff_{Wn,andTE0}|$, which can easily lead to undesirable mode coupling. Nevertheless, as long as the Y-junction length L_y is long enough, accurate and efficient, mode sorting can be achieved. Figure 4.25d shows an example of inaccurate mode sorting. When $W_n = 264$ nm, the input TE_1 mode will evolve into the TM_0 mode in port 1. Such a mode sorting must be avoided in designing devices based on asymmetric Y-junctions.

In the asymmetric Y-junction, because the effective refractive index of the input TE_0 and the TE_0 in the wide arm are always closest, the TE_0 mode sorting is determined. Here, we are mainly concerned about the input TE_1. Figure 4.26 shows the relationship between the minimum Y-junction length L_y, which achieves different mode conversion efficiency from TE_1 to TE_0 in port 2, and narrow arm width W_n.

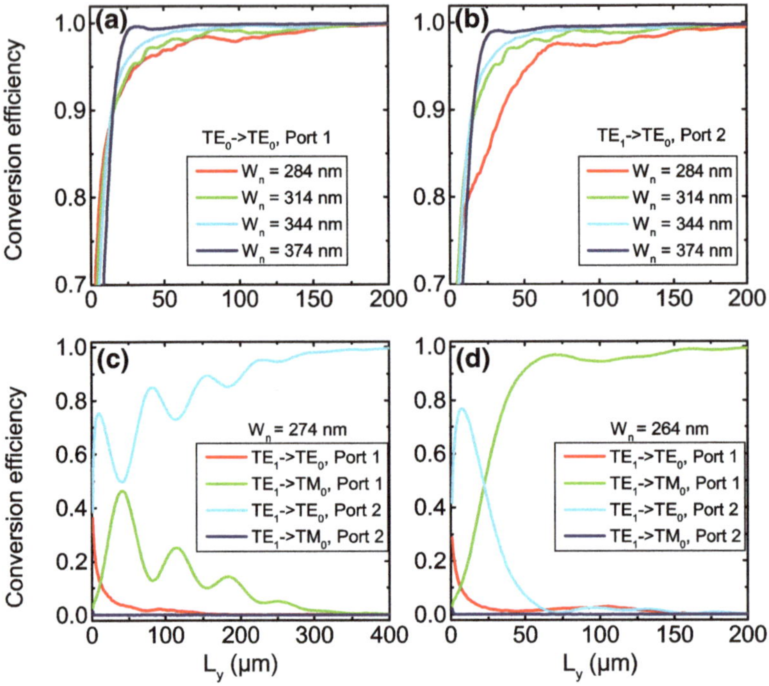

Fig. 4.25 Mode conversion efficiency in the asymmetric Y-junction for different input modes at the 1550 nm wavelength. **a**–**b** The correct mode-sorting occurs when W_n varies from 284 nm to 374 nm. **c** The undesirable mode coupling occurs when $W_n = 274$ nm. **d** The incorrect mode-sorting occurs when $W_n = 264$ nm. Reproduced from Ref. [39] by permission of Optical Society of America

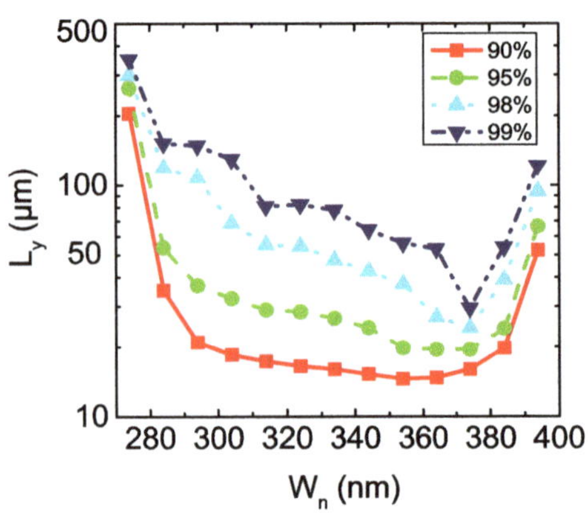

Fig. 4.26 The minimum Y-junction length required to achieve various levels of mode conversion efficiency from the TE_1 mode to the TE_0 mode in port 2 as a function of the narrow arm width W_n. Reproduced from Ref. [39] by permission of Optical Society of America

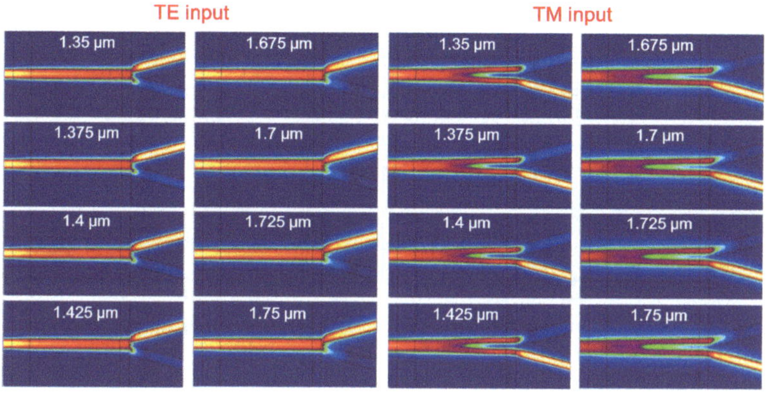

Fig. 4.27 Mode propagation in the device for different incoming polarizations at different wavelengths. Reproduced from Ref. [39] by permission of Optical Society of America

Obviously, there is a trade-off between Y-junction length L_y and the mode conversion efficiency. By increasing the length of the Y-junction, an improved mode conversion efficiency can be obtained within a large range of W_n, showing better fabrication tolerance than that of ADC-PSR. In particular, when $W_n = 374$ nm, the shortest Y-junction length L_y can be obtained for the conversion efficiency of 99%. Finally, we chose the Y-junction length $L_y = 30$ μm.

Figure 4.27 shows the propagation of different input modes in the whole device at wavelengths from 1.35 to 1.75 μm. According to the above optimization, the whole device length is 95 μm. It can be found that for TM_0 input, for such a large wavelength range, most of the power can be converted to the TE_0 mode in port 2 while the input TE_0 mode is not converted and finally output in port 1.

Figure 4.28a, b shows the wavelength dependence of mode conversion efficiency in port 1 and port 2, respectively. In addition to using the EME algorithm provided by FIMMPROP, we also used the three-dimensional FDTD simulation provided by Lumerical FDTD to carry out the simulation and verification again. In the FDTD simulation, the regions are divided into several 10 nm grids in three directions. By a comparison, it is found that the simulation results by EME and FDTD are in good agreement. When the wavelength is about 1.55 μm, the best device performance can be obtained. As the two basic components of the PSR device are broadband structures, the whole device has broadband characteristics. Within the wavelength range from 1.35 to 1.75 μm, both of these two ports have extinction ratio of 12 dB and insertion loss of less than 0.4 dB, which has a very obvious performance improvement compared to all the PSR devices reported previously. This device has potential applications in the ultra-broadband polarization diversity scheme across all communication bands.

We further analyzed the fabrication tolerance of this device, such as the waveguide width deviation Δw, the slab height error ΔH_{slab} and the refractive index error of the upper cladding material Δn_{SiO2}. Figure 4.29a, b show that when the waveguide width

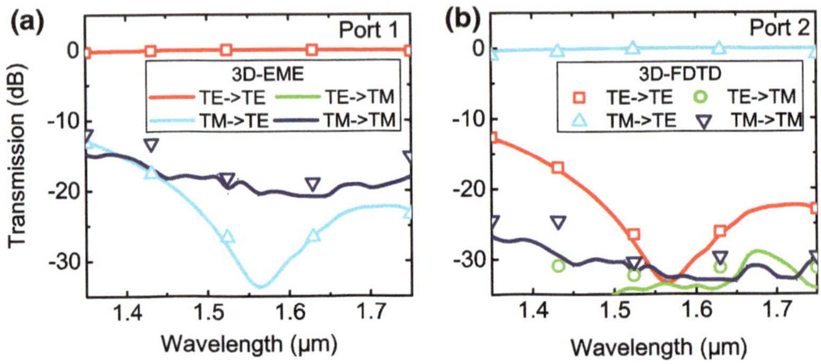

Fig. 4.28 The mode conversion efficiency as a function of the wavelength in the two output ports. The simulation is carried out by 3D-EME and 3D-FDTD methods. Reproduced from Ref. [39] by permission of Optical Society of America

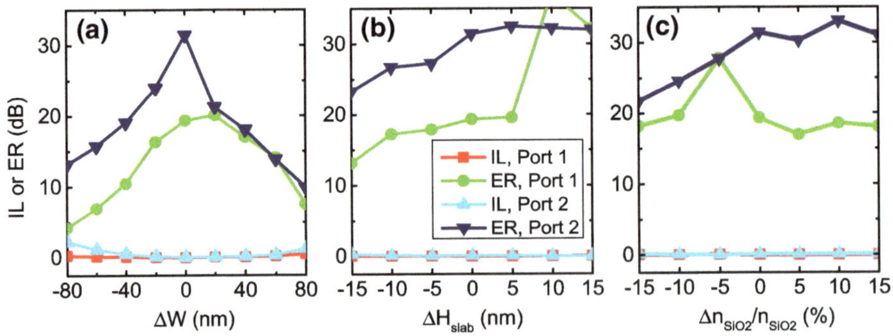

Fig. 4.29 Fabrication tolerance analysis for (**a**) the width variation Δw of the device width, (**b**) the variation ΔH_{slab} of the slab height, and (**c**) the variation $\Delta n_{SiO2}/n_{SiO2}$ of the upper-cladding refractive index. Reproduced from Ref. [39] by permission of Optical Society of America

error Δw is between -40 and 70 nm, and the slab height error ΔH_{slab} is between -15 and 15 nm, the whole device have less than 0.5 dB IL and better than 10 dB ER in these two ports. Another fabrication error occurs in the upper cladding oxidation deposition, which may lead to the refractive index of the upper cladding material is different from the lower cladding. Figure 4.29c shows that when the difference of the refractive index of the upper and lower cladding materials is $\Delta n_{SiO2}/n_{SiO2} = \pm 15\%$, that is, the refractive index of the upper cladding changes from 1.30 to 1.76, the performance of the device is still very stable. We believe that such devices can significantly improve the actual processing yield.

The discussion on how to improve the performance of mode converters and Y-junction can be referred to previous chapters. We won't explore it in this chapter.

4.4 Fabrication-Tolerant SOI PSR Based on Cascaded MMI Couplers and an Bi-Level Taper

In the previous sections, we put forward various Si-PSR designs based on the SOI rib waveguide. However, because silicon photonic devices are mainly based on strip waveguides, any rib waveguide devices need mode converters to connect the rib waveguide to the strip waveguide. Moreover, the height of the rib waveguide has great influence on the device performance, and the requirements on the fabrication accuracy are quite high. Therefore, in this section, we proposed a scheme for the realization of Si-PSR by using strip waveguides. The whole device uses a bi-level waveguide to break the symmetry of the waveguide cross-section only in the mode conversion part. The device also consists of some devices with large tolerance, such as adiabatic mode converter, multimode interference (MMI) coupler and phase shifter. These structures are more suitable to be fabricated in the CMOS platform than ADC and asymmetric Y-junction.

4.4.1 Mode Conversion Scheme in the SOI Taper

Tapered waveguide, whose width linearly increases along the light propagation direction, is one of the most important basic structures in integrated optics, and is widely used in various applications. Since the mode conversion between TM and TE is always involved in PSR devices, it is important to discuss the typical SOI tapered waveguide with SiO_2 cladding and its mode conversion mechanism for the design of a novel CMOS-compatible PSR. Here, we discuss three typical SOI tapered waveguides, as shown in Fig. 4.30.

- Figure 4.30a shows a normal tapered waveguide based on strip waveguides. It has a rectangular waveguide core and uniform upper and lower cladding, so it is symmetrical in the horizontal or vertical direction of the cross-section of the waveguide. For such a symmetrical tapered waveguide, as long as the waveguide length is sufficient, the input guided mode can keep adiabatic transmission in the waveguide and not be converted to any other modes. We give an example of a tapered waveguide with the parameters as follows: the length is 50 μm, the height of the waveguide is 0.25 μm, and the width increases from 0.50 to 0.80 μm. We used FIMMWAVE to calculate the effective refractive index of each guided mode supported by the waveguide along the propagation direction of light at 1550 nm wavelength. It can be found that there is no mode hybridization in the width range of the tapered waveguide. Even when TM_0 and TE_1 have the same effective refractive index, the different modes can be very distinguished. Therefore, the input TM_0 mode can directly pass through the tapered waveguide without any mode conversion. We used FIMMPROP to simulate the mode propagation of the TM_0 input in the tapered waveguide. The TM_0 mode does keep the polarization state unchanged and only the size of the mode field is expanded.

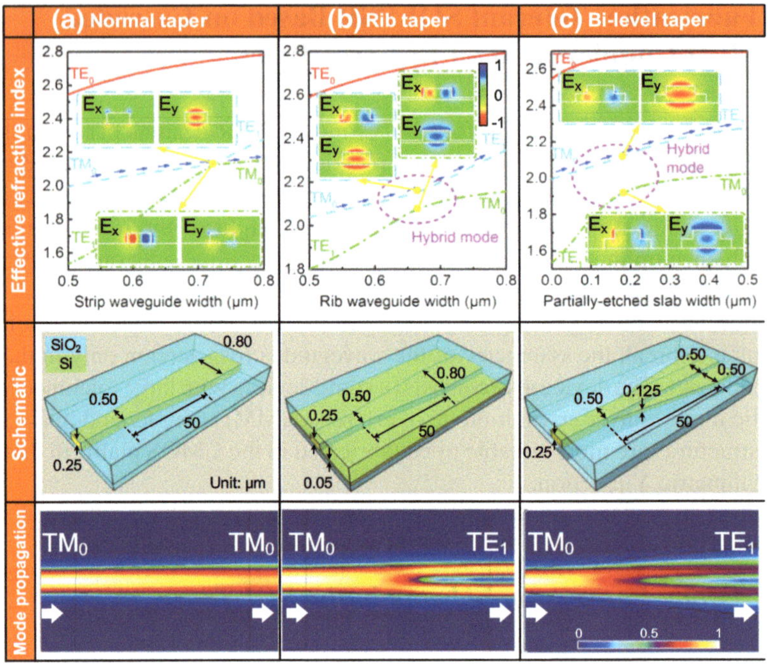

Fig. 4.30 Comparison of three kinds of SOI tapers with SiO_2 upper cladding. **a** Strip waveguide avoids mode conversion. **b** Rib waveguide supports mode conversion. **c** Bi-level waveguide supports mode conversion. Reproduced from Ref. [41] by permission of Optical Society of America

- For the tapered waveguide with asymmetric cross-section, the mode transmission in the waveguide will be much more complicated. There are many ways to make the waveguide cross-section asymmetric, such as different upper and lower cladding materials, rib waveguide or other double-etching structures. The asymmetry of waveguide cross-section may lead to indistinguishable mode of different polarization states [40]. Here, we consider two typical asymmetric tapered waveguides. Compared with the previous tapered waveguide, the waveguide core layer has a slab layer. Figure 4.30b shows a tapered mode converter based on rib waveguides. Its flat height is 0.05 μm, the width is assumed to be infinite and the width of the waveguide is increased from 0.50 to 0.80 μm linearly. Figure 4.30c shows a tapered mode converter based on bi-level taper. The width of the waveguide is constant, the height of the slab layer is 0.125 μm, and the width increases from 0 to 0.50 μm. By calculating the effective refractive index of the mode of the waveguide cross-section along the light propagation direction, it is found that when the waveguide width or the slab width increases, there will be a mode hybridization between the second modes and the third modes. The distribution of these two modes also show that the secondary components (E_x or E_y) of the mode are similar to the main components (E_y or E_x). This situation will lead to the conversion among different polarization modes. We simulated the mode propagation of input TM_0 in these two tapered waveguides. It is obvious that the input TM_0 mode will gradually evolve into the TE_1 mode. In the case of mode conversion in waveguides, a more detailed study can be found in Ref. [36].

4.4.2 *Mode Conversion in MMI Coupler*

MMI coupler is one of the basic structures in integrated optics, and is often used as optical splitter or combiner. It is based on the self-imaging principle [42]. In particular, in a multi-mode waveguide, N-fold images of the field at the input plane will be formed at the output after propagation because of the mode interference.

The input field Ψ(x, y, 0) can be expressed as the superposition of the guided modes φ_ν(x, y) as follows

$$\Psi(x, y, 0) = \sum_{\nu} c_\nu \varphi_\nu(x, y) \tag{4.7}$$

where

$$c_\nu = \frac{\int \Psi(x, y, 0) \cdot \varphi_\nu(x, y)dxdy}{\sqrt{\varphi_\nu^2(x, y)dxdy}} \tag{4.8}$$

Therefore, the MMI width should be large enough to support enough modes, otherwise some power will be lost when the field is launched into the multimode region. And the field at the position z in the propagation direction can be expressed as

$$\Psi(x, y, z) = \sum_{\nu} c_\nu \varphi_\nu(x, y) exp[j\frac{\nu(\nu + 2)\pi}{3L_\pi}] \tag{4.9}$$

where

$$L_\pi = \frac{\pi}{\beta_0 - \beta_1} \tag{4.10}$$

is the beat length of the two lowest-order modes of the multi-mode region where β_0 and β_1 are the propagation constant of the zero-order and first-order modes, respectively.

For the *general interference* where the input field enters into the multi-mode region from any x position, the field at the position of z = $3L_\pi/2$ can be expressed as

$$\Psi(x, y, \frac{3}{2}L_\pi) = \sum_{\nu\ even} c_\nu \varphi_\nu(x, y) + \sum_{\nu\ odd} (-j) \cdot c_\nu \varphi_\nu(x, y) \tag{4.11}$$

$$= \frac{1 - j}{2} \cdot \Psi(x, y, 0) + \frac{1 + j}{2} \cdot \Psi(-x, y, 0) \tag{4.12}$$

Therefore, if the input field is even symmetric ($\Psi(x, y, 0) = \Psi(-x, y, 0)$), e.g., for TE_0 mode, the field at $z = 3L_\pi/2$ will be the copy of the input field. That is

$$\Psi(x, y, \frac{3}{2}L_\pi) = \Psi(x, y, 0) \tag{4.13}$$

A special case is that the input field enters into the multimode waveguide from two symmetrical waveguides, and the input mode of the two waveguides have a phase difference of 90°. The input field in this case can be expressed as

$$\Psi(x, y, 0) = \phi(x, y, 0) + \phi(-x, y, 0) \cdot exp(j \cdot \frac{\pi}{2}) \tag{4.14}$$

Here we suppose when < 0, $\phi(x, y, 0) = 0$

According to Eqs. 4.11 and 4.14, when $z = 3L_\pi/2$, the field can be expressed as

$$\Psi(x, y, \frac{3}{2}L_\pi) = (1 + j) \cdot \phi(-x, y, 0) \tag{4.15}$$

The above formula means the input field from the two waveguides finally outputs only in an output waveguide.

Another case is called *symmetric interference*, which requires the input field to enter the multi-mode waveguide at $x = 0$. For symmetric input fields (such as TE_0), it can be reproduced at the location of $Z = 3L_\pi/4$. For the antisymmetric input field (such as TE_1), it will be divided into two parts at the same location [43]. Figure 4.31 shows the mode propagation in a MMI coupler when the inputs are symmetric field (TE_0) and antisymmetric field (TE_1) respectively.

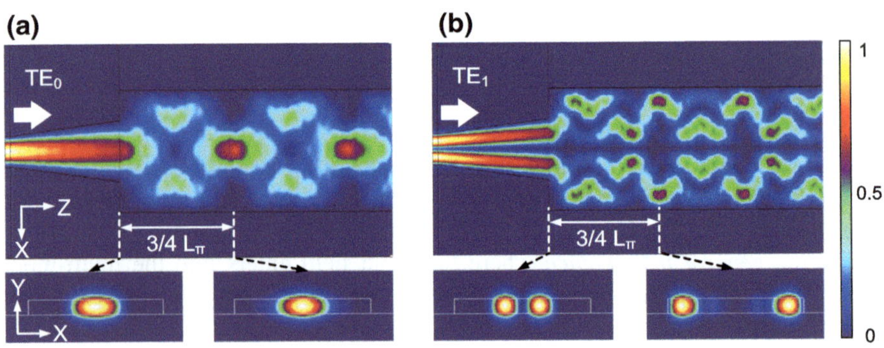

Fig. 4.31 a–b Mode propagation in the MMI coupler for the input TE_0 and TE_1 modes, respectively. Insets: the mode profile at the input plane and the position of $z = 3L_{pi}/4$. Reproduced from Ref. [41] by permission of Optical Society of America

4.4.3 Design and Optimization of the Device

Figure 4.32 shows the schematic of the proposed PSR. It has a bi-level mode converter at the front end, a 90° phase shifter in the middle, and two MMI couplers at the end. In order to meet the requirements in most silicon photonic CMOS process line, the device is designed at the SOI waveguide with 220 nm top silicon and SiO_2 upper and lower cladding. In order to break the symmetry of the waveguide cross-section to get the mode conversion between TE and TM and reduce the process difficulty, we only introduce the an extra etching process in the mode converter part, which means that the waveguide in the region has a slab layer with 90 nm height. The bi-level tapered mode converter is composed of two parts. The width of the first half of the tapered waveguide changes from W_0 to W_1, the width of the slab layer changes from 0 to W_S and the length is L_{tp1}; the width of the second half changes from W_1 to W_2, while the width of the slab layer changes from W_S to 0 and the length is L_{tp2}. The width of the two MMI couplers is W_{MMI} and the lengths are L_{MMI1} and L_{MMI2} respectively. The lengths of the input and output waveguide of the MMI coupler are L_{tp3}. The 90° phase shifter in the middle is achieved by increasing the waveguide width. The maximum waveguide width difference is dW and the length is L_{PS}. The operation principle of the whole device is described as follows:

- For the input TM_0 mode, it first converts to TE_1 in the first-order bi-level mode converter and keeps the TE_1 mode unchanged in the second-order mode converter. In order to accurately route the TE_1 mode to the TE_0 mode in port 1, we used two MMI couplers based on general interference and symmetric interference respectively. When the length of the first MMI is $L_{MMI1} = 3L_\pi/4$, the input TE_1 can be mapped into two TE_0 modes with the same intensity and 180° phase difference (see Fig. 4.31b). Then, the upper TE_0 will pass through the 90° phase shifter, so that the phase difference of the two input TE_0 of the second MMI coupler is 90°. Finally, when the length of the second MMI coupler is $L_{MMI2} = 3L_\pi/2$, according to Eq. 4.14, the two TE_0 will only be exported in port 1 with TE_0.
- For the input TE_0 mode, it will not be converted to any other guided mode in the mode converter, and then mapped to the intermediate waveguide between the first and second MMI couplers (see Fig. 4.31b), and finally output in port 2 (see Eq. 4.13).

Figure 4.33 shows the relationship between the effective refractive index of the first three guided modes in the bi-level tapered mode converter and the waveguide/slab width when the wavelength is 1550 nm. It can be found that when the waveguide width is increased from $W_0 = 0.45\ \mu m$ to $W_1 = 0.55\ \mu m$ and the slab width increases from 0 μm to $W_S = 0.50\ \mu m$, a mode hybridization region will appear for the TM_0 and TE_1 modes, which helps to realize the mode conversion between the TM_0 to TE_1 modes in the first part of the mode converter. The second part of the tapered waveguide needs to maintain the mode conversion to avoid TE_1 re-converting to TM_0 back. Therefore, we choose $W_2 = 0.85\ \mu m$ to ensure that the effective refractive index difference of the two modes at the end of the mode converter is large enough. Larger

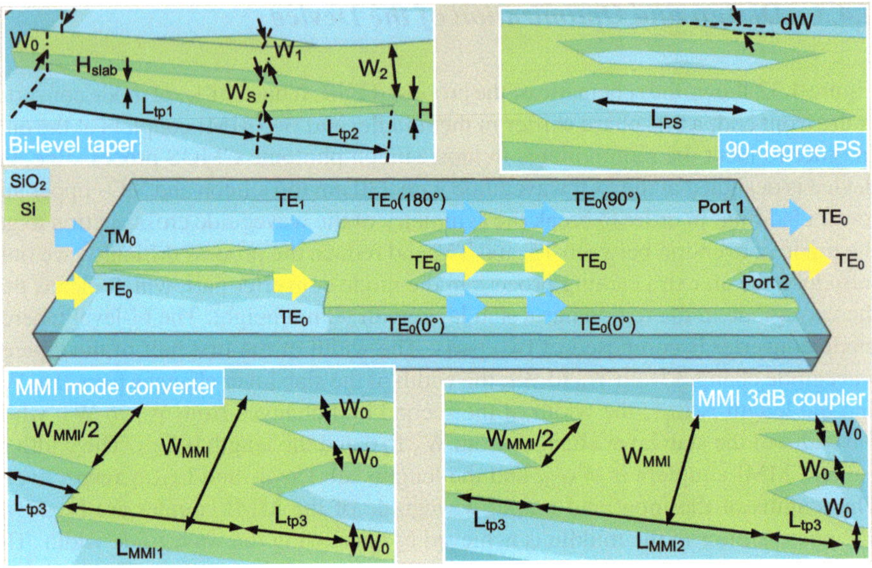

Fig. 4.32 Schematic of the proposed PSR which consists of a bi-level taper, a MMI mode converter, a 90° PS and a MMI 3 dB coupler. Reproduced from Ref. [41] by permission of Optical Society of America

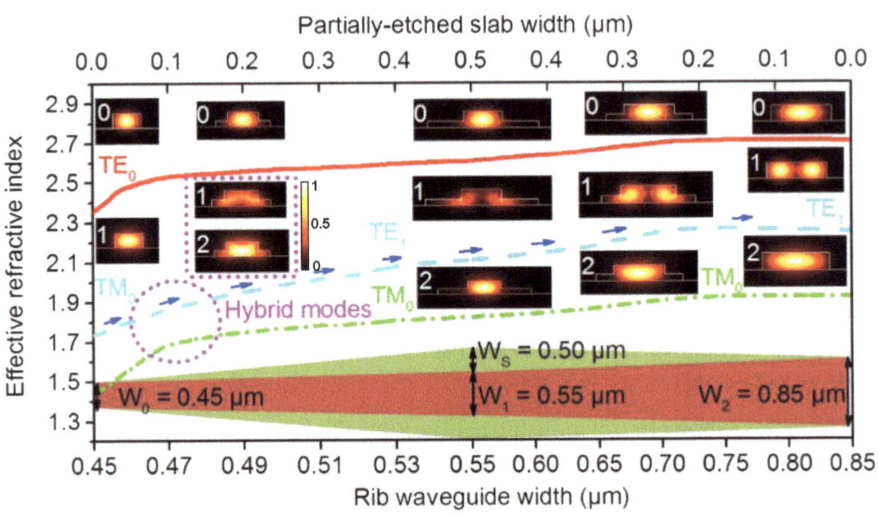

Fig. 4.33 Effective refractive indices of the first three modes in the waveguide cross-section along the bi-level taper. Insets: the profiles of the zero-, first- and second-order modes in the cross-section along this taper. Reproduced from Ref. [41] by permission of Optical Society of America

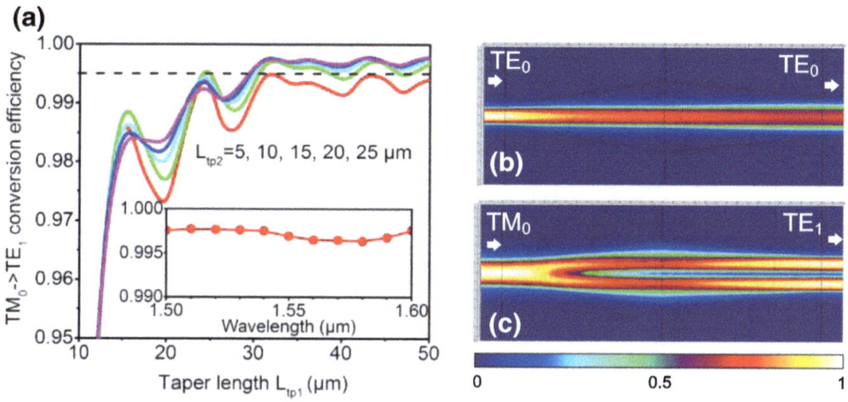

Fig. 4.34 **a** Mode conversion efficiency from TM_0 to TE_1 in the bi-level taper as a function of L_{tp1} with L_{tp2} varying from 5 μm (red) to 25 μm (pink). Inset: the wavelength dependence of the mode conversion efficiency from TM_0 to TE_1. **b–c** Mode propagation in the bi-level taper when the inputs are TE_0 and TM_0, respectively. Reproduced from Ref. [41] by permission of Optical Society of America

W_2 will lead to longer length of the mode converter, while smaller W_2 may cause unnecessary mode hybridization.

Figure 4.34a shows the mode conversion efficiency of the input TM_0 to TE_1 in the whole mode converter when different L_{tp2} is used. For different L_{tp2}, the mode conversion efficiency will increase with L_{tp1}. Here, we choose L_{tp1} = 35 μm and L_{tp1} = 20 μm to ensure that the mode conversion efficiency is more than 99.5% in the range from 1.50 to 1.60 μm. Increasing the length will further improve the mode conversion efficiency and bandwidth, but will increase the footprint of the whole device. Figure 4.34b–c show the mode propagation in the whole mode converter for different input modes, respectively. The mode conversion can be found as we expected.

Figure 4.35a shows the relationship between the mode conversion efficiency from TE_1 to TE_0 on the upper of the first MMI for different input and output waveguide lengths L_{tp3} when the MMI width W_{MMI} = 3.4 μm.When the first MMI length L_{MMI1} = ~19.9 μm, it has a high conversion efficiency of about 48.3%. At the same time, we choose the input and output waveguide length L_{tp3} = 20 μm to guarantee a low loss. Figure 4.35b shows the relationship between the phase difference between the direct waveguide and the phase shifter and the waveguide difference dW of the phase shifter when the length of the waveguide is L_{PS} = 10 μm. Compared with the phase shifter used in Refs. [44, 45], our phase shifter is broadened towards the outside, which helps to reduce the unnecessary waveguide coupling and improve the crosstalk performance of the device. In addition, it can be found that when dW = ~0.05 μm or 0.3 μm, there will be 90° phase difference. Here, we choose dW = 0.3 μm to reduce the process difficulty. For the second MMI coupler, according to the

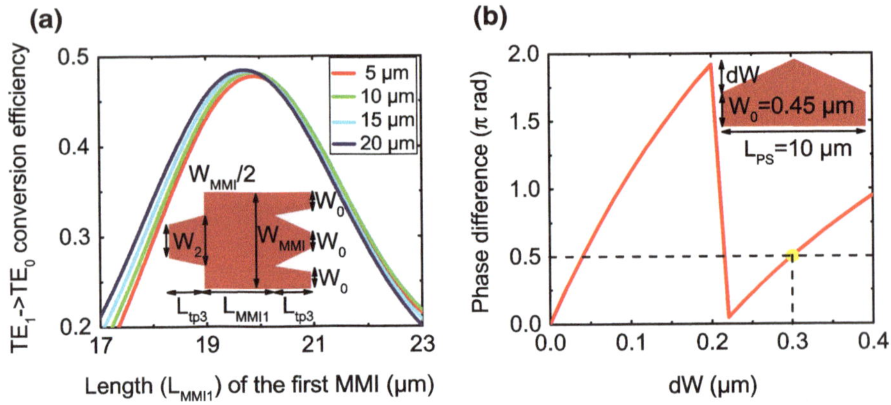

Fig. 4.35 **a** Optimization for the first MMI coupler with different input/output taper lengths (i.e., $L_{tp3} = 5, 10, 15$, and 20 μm). The MMI width W_{MMI} is chosen to be 3.4 μm here. **b** Phase difference between lights propagating through a straight waveguide and a PS with the same length $L_{PS} = 10$ μm. Reproduced from Ref. [41] by permission of Optical Society of America

operation principle of the MMI coupler, it is two times the length of the first MMI, so we choose $L_{MMI2} = \sim 39.8$ μm here.

4.4.4 Device Characterization and Fabrication Tolerance Analysis

Figure 4.36a, b show the propagation of the input TE_0 and TM_0 in the whole device respectively when the wavelength is 1550 nm. It is found that the mode propagation and conversion agrees well with the design previously, and there is no extra power coupling in the waveguide between the two MMI couplers, which proves that the phase shifter designed by us does improve the unnecessary power coupling.

Figure 4.37 shows the wavelength dependence of device loss and crosstalk performance with different input modes. When the wavelength is 1.55 μm, for TE_0 and TM_0 modes, the loss are 0.2 dB and 0.29 dB respectively, and the crosstalk between −20 dB and −20.7 dB respectively. The performance of TM_0 is more sensitive to wavelength, mainly because its mode conversion is more complicated. For example, the mode converters and phase shifters affect the conversion of the TM_0 without affecting the TE_0 mode. Nevertheless, the device maintains very good performance in the C band with less than 0.5 dB loss and less than −17 dB crosstalk in the case of two input modes. And for TE_0 input, it has acceptable performance in the range from 1.52 to 1.63 μm with less than 1 dB loss and less than −15 dB crosstalk. Since the wavelength dependence of this PSR device is mainly determined by MMI coupler, it is possible to further optimize the MMI by using some wavelength-insensitive

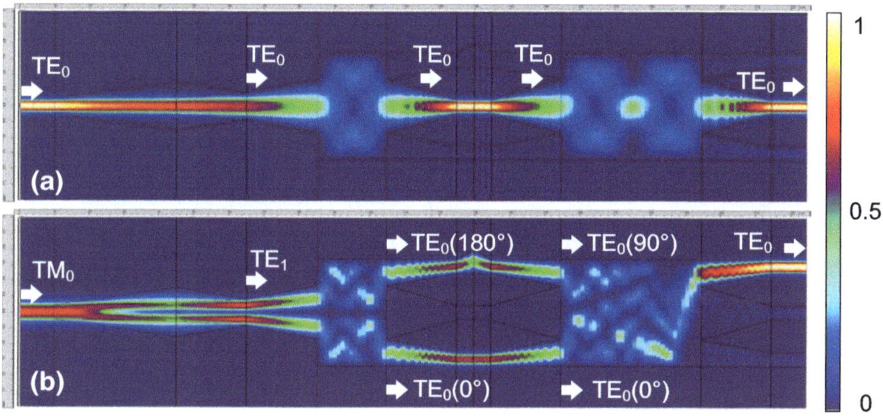

Fig. 4.36 Mode propagation in our proposed PSR at 1550 wavelength for input TE_0 and TM_0, respectively. Reproduced from Ref. [41] by permission of Optical Society of America

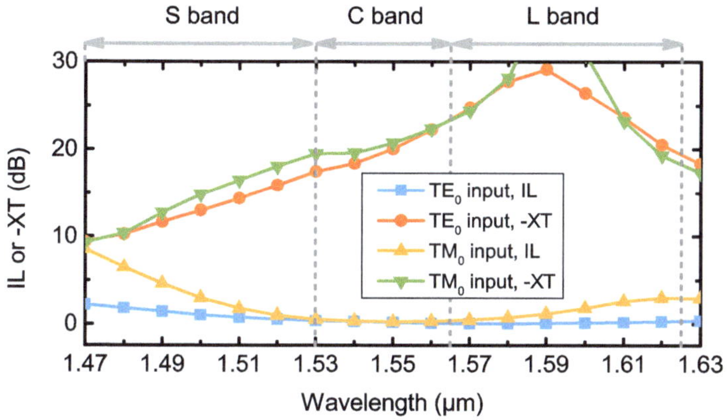

Fig. 4.37 Wavelength dependence of the PSR performance in terms of IL and XT for different input modes. Reproduced from Ref. [41] by permission of Optical Society of America

sub-wavelength structure [46]. Moreover, the crosstalk performance can be improved by cascading a polarization filter.

Figure 4.38(a1)–(a3) show the effect of the slab height error ΔH_{slab} on the device performance when the wavelengths are 1.52 μm, 1.55 μm, and 1.63 μm, respectively. Compared with the PSR based on ADC we previously reported, the performance of this PSR device has not been significantly affected even by the $\Delta H_{slab} = \pm 40$ nm. This is because the mode conversion in the bi-level taper is based on mode evolution rather than mode interference. Figure 4.38(b1)–(b3) discusses the effect of MMI width error ΔW_{MMI} on the device performance. Due to the high process tolerance of MMI, the PSR device exhibits stable performance for $\Delta W_{MMI} = 50$ nm at wavelength of 1.55 μm. At other operation wavelengths, the MMI width with

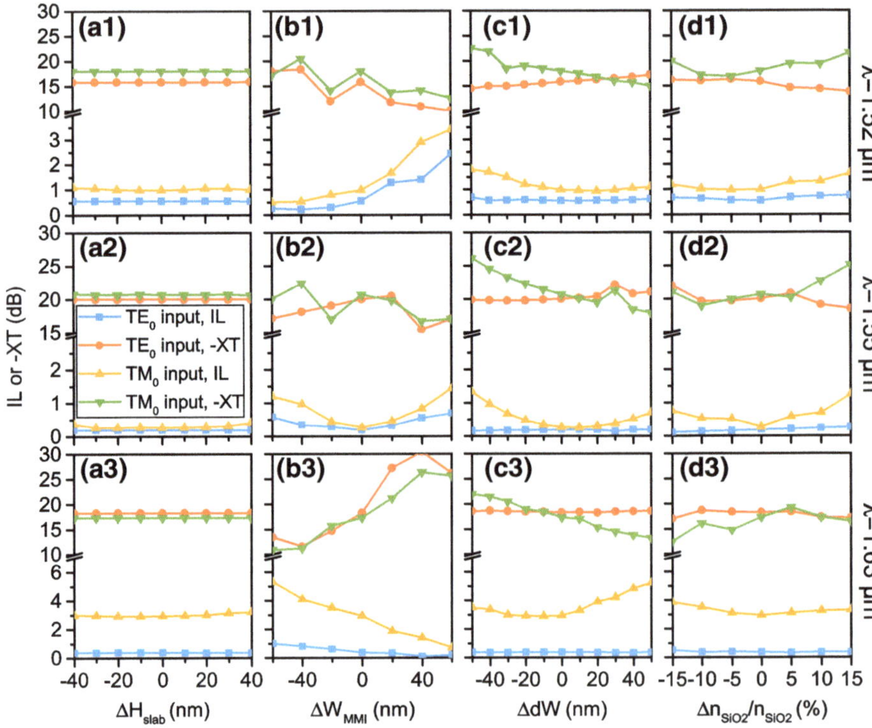

Fig. 4.38 Fabrication tolerance analysis to the deviations of (a1)–(a3) slab height ΔH_{slab} in the bi-level taper, (b1)–(b3) MMI width ΔW_{MMI}, (c1)–(c3) PS width ΔdW, and (d1)–(d3) refractive index of the upper-cladding Δn_{SiO2} at wavelengths of 1.52 μm, 1.55 μm and 1.63 μm, respectively. Reproduced from Ref. [41] by permission of Optical Society of America

optimal performance will move, for example, the smaller the wavelength, the smaller the optimal MMI width. This is mainly because the beat length of L_π will increase with the wavelength. Figure 4.38(c1)–(c3) describes the effect of ΔdW error on the performance of phase shifter. It can be found that this error will only affect in the case of TM_0 input. Nevertheless, the performance of TM_0 input is acceptable and it is also stable enough. Another process error occurs in the upper cladding oxidation deposition, which may lead to difference between the upper cladding and the lower cladding. Fortunately, the device performance is still very robust when the refractive index of the upper and lower cladding materials is $\Delta n_{SiO2}/n_{SiO2} = \pm 15\%$, that is, the refractive index of the upper cladding changes from 1.30 to 1.76. We believe that such high process tolerances can greatly improve the yield and have great advantages for realizing high-density and complex functions of photonic integrated circuits.

In addition to the loss caused by mode conversion in MMI, the scattering loss caused by the sidewall roughness of the waveguide has great effect on the total loss of the device. This loss can be improved by optimizing the process [47]. Moreover, we have also used the CMOS process to achieve a ultra-low excess loss in the MMI coupler, which also helps us to realize low-loss MMI-PSR device in the future [48].

In addition, the issue of device length is discussed. The length of the PSR device is mainly limited by the length of the mode converter and the MMI coupler. The length of the mode converter can be optimized by using a mode converter [37] with other shapes or using a multi-variable optimization technique [38]. For the MMI length, the MMI width can be carefully optimized because the MMI length is proportional to the square of the MMI width.

4.5 PSR in the Mid-Infrared

4.5.1 Si-PSR with SiN Upper Cladding at 4 μm *Wavelength*

Although the initial application of silicon photonics is the optical communication in the near infrared, researchers have extended the research of silicon photonics to the mid-infrared band in the last five years, hoping to explore different sensing applications in this wavelength range to achieve the purpose of "on-chip lab" [49–52]. Many basic devices in mid-infrared photonics have been reported, including low-loss optical waveguide [53], micro-ring resonator [54], electro-optic modulator [55], thermal optical modulator [56], photo-detector [57], arrayed waveguide grating [58], multi-mode interference coupler [59], grating coupler [60] and so on. However, the refractive index difference between Si and SiO_2 is very large, which leads to strong structural birefringence. For example, the effective index difference of different guided modes is very large. Such a problem exists not only in the optical communication band, but also in the mid-infrared band. Although the polarization diversity scheme can solve such problems, all the previous reports are aimed at the optical communication band. Up to now, the polarization diversity scheme in the mid-infrared band has not been studied. This undoubtedly limits the development of mid-infrared silicon photonics.

In order to realize the polarization diversity scheme of the silicon photonics in the mid-infrared band, we propose a PSR with the assisted mode converter of tapered waveguide and an mode-sorting asymmetrical Y-junction. The PSR based on asymmetric Y-junction can be more suitable for sensing applications because of its large operation bandwidth. Compared with the near infrared photonics, a large difference in the mid-infrared band is that SiO_2 has a strong absorption of light, making it unsuitable to be a cladding material. Lin et al. reported an integrated platform based on SiN waveguides, and tested the absorption coefficient of SiN which is very low when the wavelength is less than 8 μm [61, 62]. Therefore, we use another CMOS-compatible material SiN as the cladding material for PSR, thereby reducing the absorption loss when the wavelength is greater than 4 μm. Moreover, we also studied the SiO_2 and SiN as the upper and lower cladding PSR, and their performance at shorter wavelengths (~3.3 μm) and longer wavelengths (~6.9 μm).

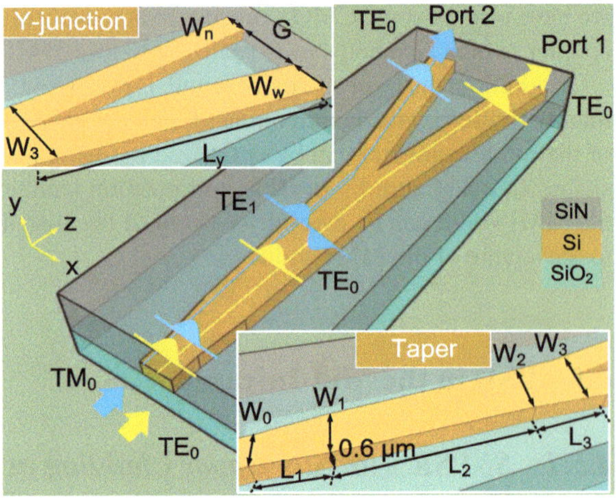

Fig. 4.39 Schematic of the proposed PSR device which consists of a three-stage taper and a mode-sorting asymmetric Y-junction for mid-infrared applications. Reproduced from Ref. [66] by permission of Optical Society of America

Figure 4.39 shows the schematic of our proposed PSR devices for mid-infrared applications. It is designed on a strip waveguide with top silicon thickness of 220 nm and upper cladding of SiN. Although Si has a transparent window of up to 8 μm wavelength, the optical loss of SiO_2 will suddenly increase at wavelength larger than 3.6 μm [63]. In order to reduce the absorption loss caused by SiO_2, we used a thicker Si layer (thickness H = 600 nm) so that the field intensity will be closer to 0 at the Si/SiO_2 interface [59]. Moreover, we propose to use SiN as the upper cladding of PSR devices which is CMOS-compatible and has low-loss window up to 7 μm wavelength. Because the refractive index of SiN is larger than that of SiO_2, more light will be confined in SiN layer, which will help to reduce the absorption loss. In addition, after using the SiN cladding, the symmetry of the cross-section of the waveguide is broken naturally as previously reported [6, 64], so that there is no need to use rib waveguide or bi-level waveguide in the mode conversion part. Thus the whole PSR device needs only one step to etch, which can obviously improve the yield of the device.

The PSR device consists of a three-order linear mode converter and an mode-sorting asymmetric Y-junction. We design and optimize the three-order linear mode converter to ensure that the input TM_0 mode is slowly converted to TE_1 mode without changing the input TE_0 mode. If the design parameters of the asymmetric Y-junction are carefully selected, such as the width of the arm waveguide and the Y-junction length, the TE_1 mode will be converted to TE_0 and output in port 2 while the TE_0 mode output in port 1.

Figure 4.40 shows the relationship between the effective refractive index of the first three guided modes of the SOI strip waveguide with the SiN upper cladding and the SiO_2 bottom cladding and the width of the waveguide when the wavelength is 4 μm. In the simulation, we use the refractive index of Si_3N_4 for the SiN cladding [65], and analyze the refractive index variation as a process error in the next part. We choose the input waveguide width W_0 = 1.85 μm to guarantee single-mode transmission.

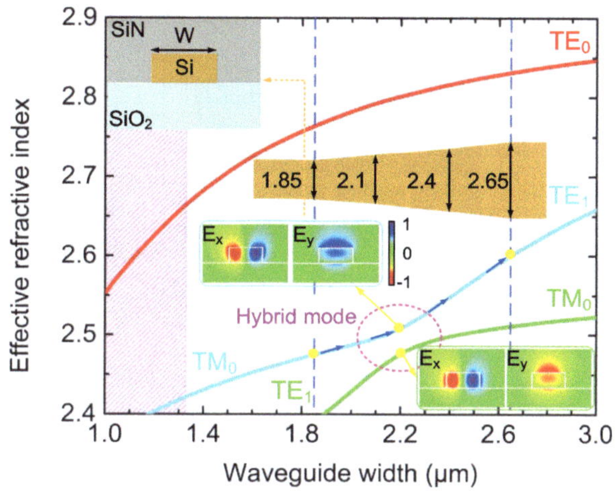

Fig. 4.40 Effective refractive indices of the first three modes in a Si waveguide with SiN upper-cladding and SiO_2 lower-cladding at 4.0 μm wavelength. Reproduced from Ref. [66] by permission of Optical Society of America

It can be found that when the width of the waveguide increases from $W_1 = 2.1$ μm to $W_2 = 2.4$ μm, the two modes of TM_0 and TE_1 will have a hybrid mode region. These two hybrid modes are not purely polarized modes. Based on the field distribution of the two mode, it is found that secondary components (E_x or E_y) are similar to the secondary components (E_y or E_x), indicating that the hybrid mode has the characteristics of the two modes of TM_0 and TE_1. For example, for the second guided mode (that is, the first hybrid mode), it has an extra part of TE_1 and will increase with the waveguide width. This leads to a discrete part in the refractive index curve of TM_0 and the conversion from TM_0 mode to TE_1 mode. After the mode conversion of the TM_0 to TE_1 is achieved in the second part of the mode converter, it needs to keep the mode conversion unchanged and avoid the TE_1 mode converted to TM_0 mode in the third part of the waveguide. Therefore, we choose the width of the waveguide $W_3 = 2.65$ μm at the end to ensure that the difference between the effective refractive index of TE_1 and TM_0 is large enough. Larger W_3 requires larger mode converter length to reduce mode conversion loss. When the asymmetric Y-junction is designed, the widths of the two arm waveguides satisfy the relationship $W_n + W_w = W_3$, where W_n and W_w are the widths of the narrow arm waveguide and the wide arm waveguide respectively. According to the principle of mode sorting, it is found that the range of W_n(pink shadow marking) is very large. However, when the W_n is further reduced, the limiting on the light is reduced, and the mode field in the Si waveguide will gradually enter the SiO_2 cladding, causing the absorption loss to increase.

Figure 4.41a shows the relationship between the conversion efficiency from TM_0 to TM_0/TE_1 and L_2 respectively when L_1 changes from 10 to 60 μm. Here, $L_3 = L_1(W_3\text{-}W_2) / (W_1\text{-}W_0)$. The ripples in the curve may be caused by some mode interference or reflection, which can be improved by increasing L_1. Here, we choose $L_1 = 60$ μm and $L_2 = 150$ μm to achieve mode conversion efficiency higher than 99%. Longer waveguides may improve mode conversion efficiency and bandwidth perfor-

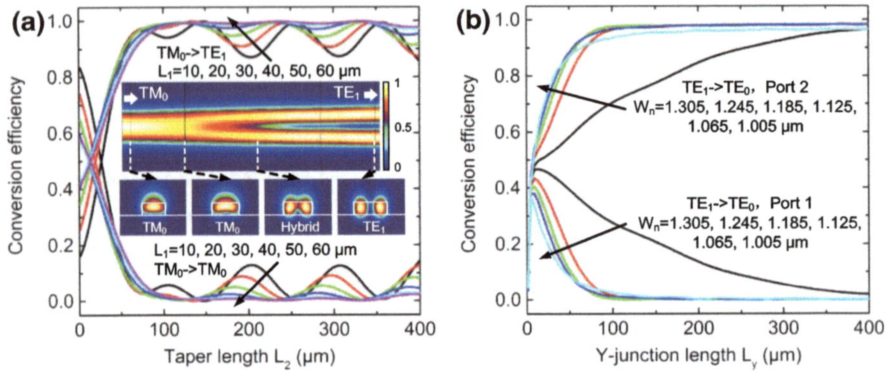

Fig. 4.41 **a** Mode conversion efficiency for the input TM_0 mode as a function of L_2 for L_1 varying from 10 μm (black) to 60 μm (pink). **b** Mode conversion efficiency in the Y-junction for the TE_1 input with different widths of the narrow arm W_n changing from 1.305 μm (black) to 1.005 μm (light blue). All the simulations were performed at 4.0 μm wavelength. Reproduced from Ref. [66] by permission of Optical Society of America

mance, but will increase the device area. The inner graph of Fig. 4.41a shows the propagation of the input TM_0 mode in the mode converter. It can be found that mode conversion is going as we expected, first evolved into a hybrid mode and then output as TE_1 mode.

Figure 4.41b shows the relationship between the conversion efficiency of TE_1 mode to the other modes and the Y-junction length with different narrow arm waveguide width W_n. When W_n changes in a large range from 1.305 to 1.005 μm, efficient and correct mode sorting can be achieved. And the mode conversion efficiency will increase with Y-junction length L_y. A high mode conversion efficiency can be obtained in such a large W_n range, which implies that the PSR device has great fabrication tolerance. Here, we choose $W_n = 1.245$ μm and $L_y = 200$ μm.

Figure 4.42a, b show the propagation of different input modes in the whole device respectively when the wavelength is 4 μm. It can be found that the input TM_0 will evolve into TE_1 in the three-order linear mode converter and output with TE_0 in port 2, and the input TE_0 directly pass through the mode converter and then output in port 1. This device has a length of 470 μm, so when the pulse signal passes through it, there is probably some dispersion. However, the method of reducing the device length introduced in the previous chapters can still be used for improvement.

Figure 4.42c shows the wavelength dependance of loss, crosstalk and extinction ratio. For the TE_0 input, the PSR device has loss of less than 0.06 dB and crosstalk better than −28 dB at wavelengths from 4 μm to 4.4 μm. The corresponding extinction ratio is always less than −40 dB because of the efficient mode-sorting effect. Here, we did not extend the coordinate system to below −40 dB. This is because the accuracy of the simulation is not guaranteed, and the absolute value is not accurate

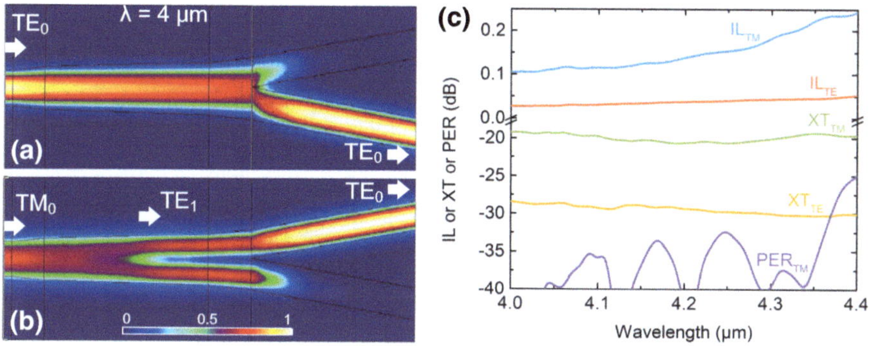

Fig. 4.42 **a–b** Mode propagation in the overall PSR device for the input TE_0 and TM_0 modes at 4.0 μm wavelength, respectively. **c** Wavelength dependence of the PSR performance in terms of IL, XT and PER. Reproduced from Ref. [66] by permission of Optical Society of America

enough. So we can only give a simple range of values. For the input TM_0, the device performance will be reduced due to more complex mode conversion. Nevertheless, in this 400 nm wavelength range, the loss is still less than 0.25 dB and the crosstalk is less than −18 dB.

We replace the SiO_2 absorption coefficient into the previous model [65] and calculate the absorption loss of the entire PSR device due to the absorption of SiO_2. It can be found that when the wavelength is between 4 μm and 4.4 μm, the absorption loss of TE_0 and TM_0 are less than 0.13 dB and 0.17 dB respectively.

We further analyze the process tolerance of the device, such as the waveguide width error ΔW, the height error ΔH and the refractive index error Δn_{SiN} of the upper cladding material, as shown in Fig. 4.43a–c, respectively. For small waveguide width error, the device performance in two cases will not be significantly affected, which is mainly due to the large waveguide width at this time. It can be found that when the waveguide width error $\Delta W = \pm 100$ nm, there is only 0.1 dB loss variation. For the other two process errors, the device with TM_0 input is more sensitive than the TE_0 input. For the input TE_0 mode, the performance of the device is very stable, because there is no mode conversion in the mode converter, and the TE_0 will always output in the wide arm waveguide. For the input TM_0 mode, the two process errors will destroy the mode conversion to the TE_1 in the mode converter, and cause some unwanted modes at the output waveguides of the Y-junction. In addition, the PSR device maintains excellent performance in a considerable range of error (i.e. $\Delta H = \pm 80$ nm and $\Delta n_{SiN}/n_{SiN} = \pm 10\%$), such as loss less than 0.4 dB, and crosstalk less than −12 dB. We note that this device is very robust for the refractive index error of the SiN cladding material, which is very important, because the refractive index of the SiN material changes obviously under the deposition condition.

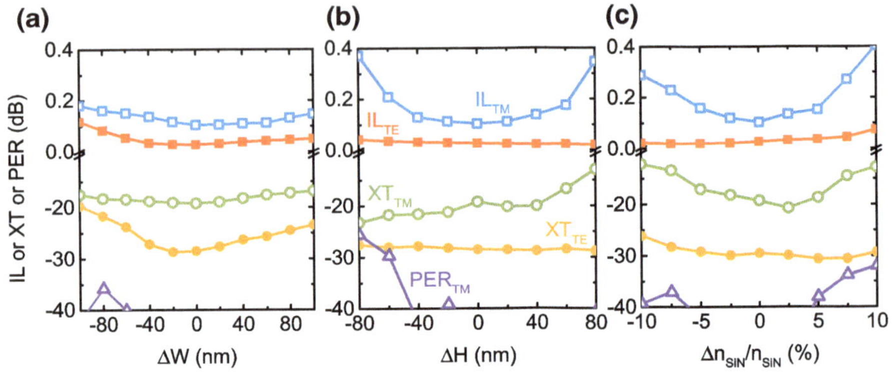

Fig. 4.43 **a–c** Analysis of fabrication tolerance with respect to the parameter deviations in waveguide width ΔW, height ΔH and refractive index of the upper-cladding Δn_{SiN}, respectively. The simulation wavelength is 4.0 μm. Reproduced from Ref. [66] by permission of Optical Society of America

4.5.2 *PSR Devices Operating at Other Mid-IR Wavelengths Using SiO_2 or SiN Cladding*

For PSR devices working in a shorter mid-infrared wavelength range (such as less than 4 μm), SiO_2 is also an upper-cladding material which has acceptable absorption loss [63]. Recently, some silicon photonic devices with SiO_2 upper and lower cladding have been reported in the mid-infrared band. It has a relatively low waveguide loss [58], which proves the feasibility of realizing the PSR devices using the SiO_2 cladding. Figure 4.44 shows a PSR device based on a bi-level tapered mode converter and an asymmetrical Y-junction with upper and lower cladding of SiO_2. Since the waveguide has the same upper and lower cladding material at this time, we use a bi-level waveguide to break the symmetry of the waveguide cross-section, thus realizing the mode conversion from TM to TE. By using the previous method, we designed and optimized the mode converter and the Y-junction at 3.3 μm wavelength. The simulation result show that for different input modes, the loss is less than 0.1 dB and the crosstalk is less than −30 dB. Figure 4.44b, c show the propagation of TE_0 and TM_0 in the device at wavelengths of 3.3 μm, respectively. You can find the mode conversion as we expected.

In order to completely remove the loss caused by SiO_2 absorption, SiN can also be used as upper and lower cladding materials. Because it is difficult to obtain Si on SiN crystal, making such a structure is challenging. But we can still start with the SOI wafer and define the waveguide structure through suitable lithography technology. After depositing a SiN cladding, it can be opened in the SiN cladding near the device. Because the etching rate of the hydrofluoric acid (HF) for SiO_2 is much higher than that of Si and SiN, the wafer can be immersed into the HF to remove SiO_2 at the bottom [54]. In this way, Si devices will be separated from the Si substrate, but will

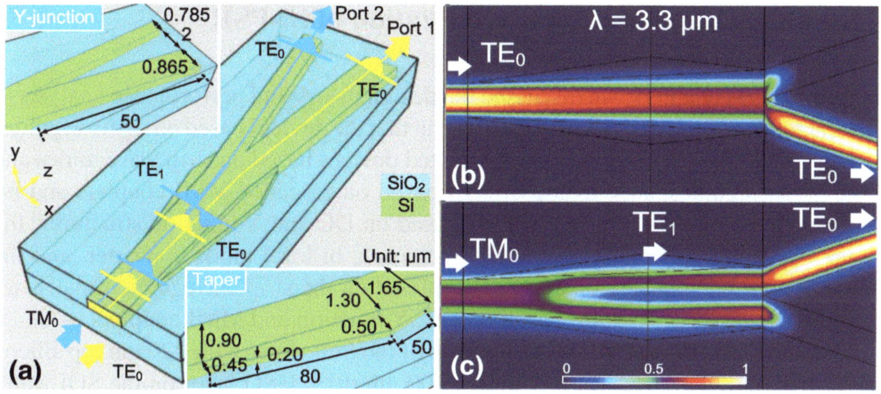

Fig. 4.44 The PSR device with SiO_2 upper- and lower-claddings. **a** Schematic of the device based on an asymmetric Y-junction and a bi-level mode-conversion taper at shorter wavelengths. **b–c** Mode propagation in the overall device at 3.3 μm wavelength for the TE_0 and TM_0 inputs, respectively. Reproduced from Ref. [66] by permission of Optical Society of America

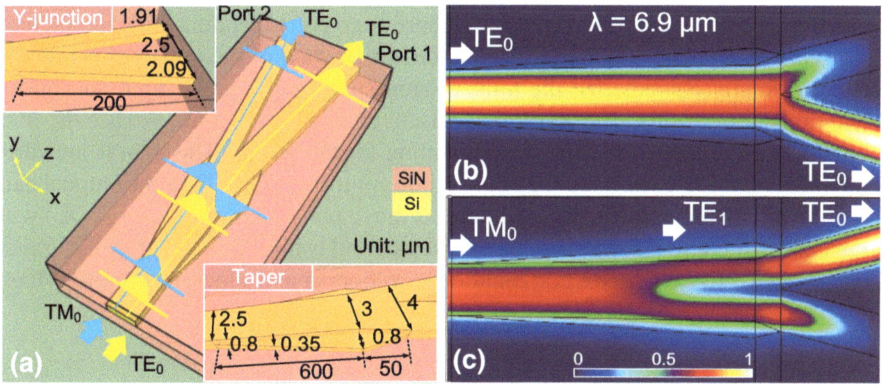

Fig. 4.45 The PSR device with SiN upper- and lower-claddings. **a** Schematic of the device based on an asymmetric Y-junction and a bi-level mode-conversion taper. **b–c** Mode propagation in the overall device at 6.9 μm wavelength for TE_0 and TM_0 inputs, respectively. Reproduced from Ref. [66] by permission of Optical Society of America

be fixed on the SiN cladding. Chemical vapor deposition of SiN can fill the gap between Si device and Si substrate, so PSR devices completely wrapped by SiN can be obtained. Figure 4.45b, c show the propagation of TE_0 and TM_0 in the entire device when the wavelength is 6.9 μm, proving that the mode conversion is efficient and accurate. For the applications at longer wavelengths, in order to further improve the absorption loss, it can be realized by using the air cladding or the etching oxide layer to form a suspension structure.

4.6 Fabrication and Characterization of Si-PSR

In the previous chapters, we discussed the design theory and performance analysis of several types of Si-PSR devices. Considering that the devices based on rib waveguides is sensitive to the process, we only fabricated devices based on strip waveguides and bi-level waveguides, such as the Si-PSR based on cascaded MMI couplers and bi-level mode converters. For the Si-PSR based on DC and Y-junction structures, the input TM_0 mode is converted to TE_1 by using a bi-level mode converter, and the subsequent mode conversion can be completed on the strip waveguide. Finally, we processed a series of samples based on DC, Y-junction and MMI couplers.

We fabricated these PSR devices on top of the SOI wafer with 220 nm top silicon and 2 μm buried oxide. First, we spin a 100 nm thick HSQ layer on the SOI wafer as negative glue for electron beam exposure. Secondly, the pattern is defined by electron beam exposure. Then, the Si layer is removed by dry etching. Finally, a 2 μm thick SiO_2 upper layer is deposited. Figures 4.46 and 4.47 show the microscope pictures of Si-PSR samples based on DC and MMI structures, respectively. Because the Y-junction is relatively long, the details of the device can not be seen clearly by using the visible optical microscope and there is no corresponding enlarged picture.

For the PSR test, we must tune the polarization state of the optical signal at the end, and finally characterize the transmission spectrum of eight input and output conditions in two ports. The test link we use is shown in Fig. 4.48. We use the fiber coupling for the input and free-space coupling for the output. The light is launched into the device (DUT) through the polarization controller (PC). Then the output beam is calibrated by the object lens (OBJ), enters into a polarization analyzer (POL) with high extinction ratio through a range of distance and is finally received by the optical power-meter. Here we used a combination of a chopper and a lock-in amplifier (LIA)

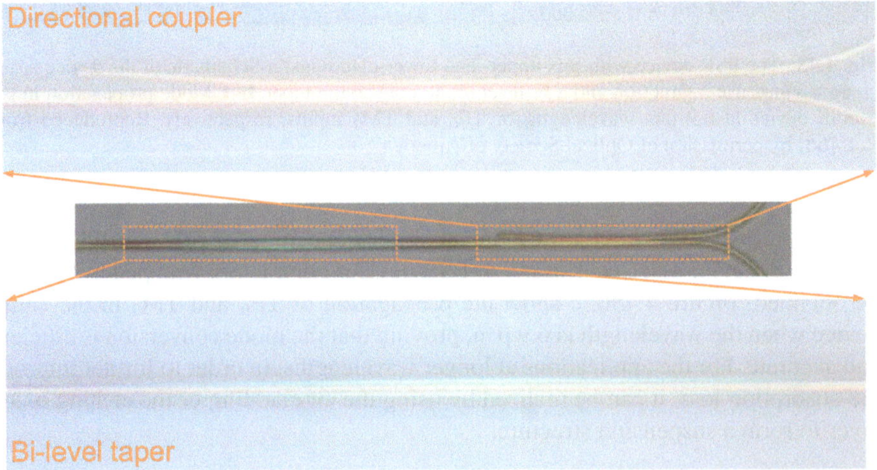

Fig. 4.46 Microscoped picture of Si-PSR sample based on DC and bi-level mode converter

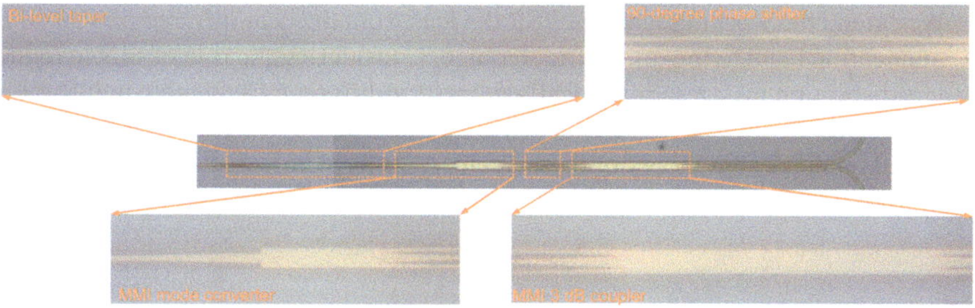

Fig. 4.47 Microscoped picture of Si-PSR sample based on cascaded MMI couplers and bi-level mode converter

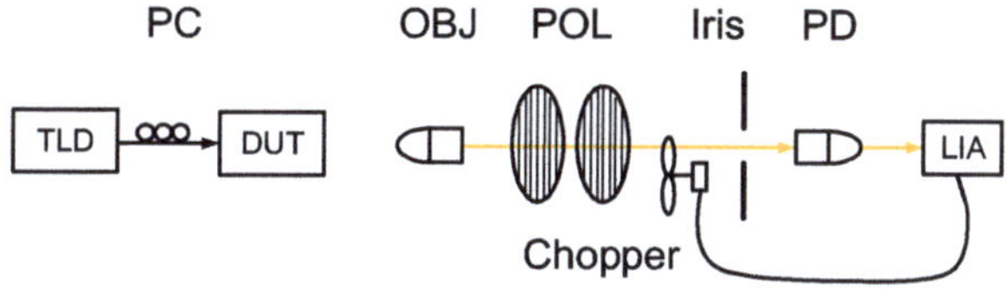

Fig. 4.48 Test set-up for characterizing PSR, PBS and PR devices

to compress the reflected light, and used a variable aperture to compress the output light with the slab mode. The specific test results will be reported in subsequent investigation.

4.7 Conclusion

In this chapter, we proposed a series of polarization splitter-rotators for polarization diversity scheme in silicon photonics. Specifically,

1. We designed an ultra-compact PSR device based on rib directional coupler with SiO_2 upper cladding. The coupling length is only 9 μm, and the total length is only 25 μm.
2. We used cascaded or assisted mode converters to improve the PSR devices based on rib directional coupler. The device performance and the process tolerance are improved relatively.
3. We proposed a ultra-broadband PSR device by using the three-order linear mode converter and the mode-sorting asymmetric Y-junction. The device has a working range of up to 400 nm wavelength which is large enough to cover all optical communication bands.
4. We used some cascaded MMI couplers, a bi-level mode converter and a 90° phase shifter to design a PSR device with high fabrication tolerance. This device is highly tolerant to various process errors, and can meet the requirements under the CMOS process conditions.

5. We extend the working range of PSR devices from the near infrared optical communication band to the mid-infrared range. Considering the role of SiN in reducing absorption loss, a series of PSR devices for different wavelength applications are proposed.
6. Based on the design theory of these devices, we have processed a series of device samples and built a test set-up.

References

1. Wang Z, Dai D, He S (2007) Polarization-insensitive ultrasmall microring resonator design based on optimized Si sandwich nanowires. IEEE Photonics Technol Lett 19:1580–1582
2. Lang T, He JJ, Kuang JG, He S (2007) Birefringence compensated AWG demultiplexer with angled star couplers. Opt Express 15:15022–15028
3. Lim ST, Png CE, Ong EA, Ang YL (2007) Single mode, polarization-independent submicron silicon waveguides based on geometrical adjustments. Opt Express 15:11061–11072
4. Barwicz T, Watts MR, Popović MA, Rakich PT, Socci L, Kärtner FX, Ippen EP, Smith HI (2007) Polarization-transparent microphotonic devices in the strong confinement limit. Nat Photonics 1:57–60
5. Watts MR, Haus HA (2005) Integrated mode-evolution-based polarization rotators. Opt Lett 30:138–140
6. Chen L, Doerr CR, Chen YK (2011) Compact polarization rotator on silicon for polarization-diversified circuits. Opt Lett 36:469–471
7. Zhang H, Das S, Zhang J, Huang Y, Li C, Chen S, Zhou H, Yu M, Guo-Qiang Lo P, Thong JT (2012) Efficient and broadband polarization rotator using horizontal slot waveguide for silicon photonics. Appl Phys Lett 101:021105
8. Komatsu MA, Saitoh K, Koshiba M (2012) Compact polarization rotator based on surface plasmon polariton with low insertion loss. IEEE Photonics J 4:707–714
9. Leung DMH, Rahman BMA, Grattan KTV (2011) Numerical analysis of asymmetric silicon nanowire waveguide as compact polarization rotator. IEEE Photonics J 3:381–389
10. Deng H, Yevick DO, Brooks C, Jessop PE (2006) Fabrication tolerance of asymmetric silicon-on-insulator polarization rotators. J Opt Soc Am A 23:1741–1745
11. Wang Z, Dai D (2008) Ultrasmall Si-nanowire-based polarization rotator. J Opt Soc Am B 25:747–753
12. Cao T, Chen S, Fei Y, Zhang L, Xu QY (2013) Ultra-compact and fabrication-tolerant polarization rotator based on a bend asymmetric-slab waveguide. Appl Opt 52:990–996
13. Gao L, Huo Y, Harris JS, Zhou Z (2013) Ultra-compact and low-loss polarization rotator based on asymmetric hybrid plasmonic waveguide. IEEE Photonics Technol Lett 25:2081–2084
14. Deng H, Yevick DO, Brooks C, Jessop PE (2005) Design rules for slanted-angle polarization rotators. J Lightwave Technol 23:432–445
15. Deng H, Yevick DO, Chaudhuri SK (2005) Bending characteristics of asymmetric SOI polarization rotators. IEEE Photonics Technol Lett 17:2113–2115
16. Aamer M, Gutierrez AM, Brimont A, Vermeulen D, Roelkens G, Fedeli JM, Hakansson A, Sanchis P (2012) CMOS compatible silicon-on-insulator polarization rotator based on symmetry breaking of the waveguide cross section. IEEE Photonics Technol Lett 24:2031–2034
17. Vermeulen D, Verheyen P, Absil P, Bogaerts W, Van Thourhout D, Roelkens G (2012) Silicon-on-insulator polarization rotator based on a symmetry breaking silicon overlay. IEEE Photonics Technol Lett 24:482–484
18. Zhang J, Yu M, Lo G, Kwong DL (2010) Silicon waveguide-based mode-evolution polarization rotator. In: Silicon photonics and photonic integrated circuits II, vol 7719

19. Zhang J, Zhu S, Zhang H, Chen S, Lo GQ, Kwong DL (2011) An ultracompact surface plasmon polariton-effect-based polarization rotator. IEEE Photonics Technol Lett 23:1606–1608
20. Fei Y, Zhang L, Cao T, Cao Y, Chen S (2013) High efficiency broadband polarization converter based on tapered slot waveguide. IEEE Photonics Technol Lett 25:879–881
21. Liu L, Ding Y, Yvind K, Hvam JM (2011) Silicon-on-insulator polarization splitting and rotating device for polarization diversity circuits. Opt Express 19:12646–12651
22. Ding Y, Liu L, Peucheret C, Xu J, Ou H, Yvind K, Zhang X, Huang D (2012) Towards polarization diversity on the SOI platform with simple fabrication process. IEEE Photonics Technol Lett 23:1808–1810
23. Ding Y, Liu L, Peucheret C, Ou H (2012) Fabrication tolerant polarization splitter and rotator based on a tapered directional coupler. Opt Express 20:20021–20027
24. Ding Y, Ou H, Peucheret C (2013) Wideband polarization splitter and rotator with large fabrication tolerance and simple fabrication process. Opt Lett 38:1227–1229
25. Ding Y, Huang B, Ou H, Da Ros F, Peucheret C (2013) Polarization diversity DPSK demodulator on the silicon-on-insulator platform with simple fabrication. Opt Express 21:7828–7834
26. Dai D, Bowers JE (2011) Novel concept for ultracompact polarization splitter-rotator based on silicon nanowires. Opt Express 19:10940–10949
27. Fei Y, Zhang L, Cao T, Cao Y, Chen S (2012) Ultracompact polarization splitter-rotator based on an asymmetric directional coupler. Appl Opt 51:8257–8261
28. Sacher WD, Barwicz T, Taylor BJ, Poon JK (2014) Polarization rotator-splitters in standard active silicon photonics platforms. Opt Express 22:3777–3786
29. Guan H, Novack A, Streshinsky M, Shi R, Fang Q, Lim AEJ, Lo GQ, Baehr-Jones T, Hochberg M (2014) CMOS-compatible highly efficient polarization splitter and rotator based on a double-etched directional coupler. Opt Express 22:2489–2496
30. Guan H, Novack A, Streshinsky M, Shi R, Liu Y, Fang Q, Lo GQ, Baehr-Jones T, Hochberg M (2014) High-efficiency low-crosstalk 1310-nm polarization splitter and rotator. IEEE Photonics Technol Lett 26:925–928
31. Guan H, Fang Q, Lo GQ, Bergman K (2015) High-efficiency biwavelength polarization splitter-rotator on the SOI platform. IEEE Photonics Technol Lett 27:518–521
32. Xiong Y, Xu DX, Schmid JH, Cheben P, Janz S, Winnie NY (2014) Fabrication tolerant and broadband polarization splitter and rotator based on a taper-etched directional coupler. Opt Express 22:17458–17465
33. Xiong Y, Wangemert-Prez JG, Xu DX, Schmid JH, Cheben P, Winnie NY (2014) Polarization splitter and rotator with subwavelength grating for enhanced fabrication tolerance. Opt Lett 39:6931–6934
34. Yin M, Deng Q, Li Y, Wang X, Li H (2015) Ultrashort and low-loss polarization rotators utilizing hybrid plasmonic-dielectric couplers. IEEE Photonics Technol Lett 27:229–232
35. Wang J, Niu B, Sheng Z, Wu A, Wang X, Zou S, Qi M, Gan F (2014) Design of a SiO_2 top-cladding and compact polarization splitter-rotator based on a rib directional coupler. Opt Express 22:4137–4143
36. Dai D, Tang Y, Bowers JE (2012) Mode conversion in tapered submicron silicon ridge optical waveguides. Opt Express 20:13425–13439
37. Luyssaert B, Bienstman P, Vandersteegen P, Dumon P, Baets R (2005) Efficient nonadiabatic planar waveguide tapers. J Lightwave Technol 23:2462–2468
38. Zou J, Yu Y, Ye M, Liu L, Deng S, Xu X, Zhang X (2014) Short and efficient mode-size converter designed by segmented-stepwise method. Opt Lett 39:6273–6276
39. Wang J, Niu B, Sheng Z, Wu A, Li W, Wang X, Zou S, Qi M, Gan F (2014) Novel ultra-broadband polarization splitter-rotator based on mode-evolution tapers and a mode-sorting asymmetric Y-junction. Opt Express 22:13565–13571
40. Vermeulen D, Van Acoleyen K, Ghosh S, De Cort W, Yebo NA, Hallynck E, De Vos K, Debackere P, Dumon P, Bogaerts W, Roelkens G (2010) Efficient tapering to the fundamental quasi-TM mode in asymmetrical waveguides. In: 15th European conference on integrated optics

41. Wang J, Qi M, Xuan Y, Huang H, Li Y, Li M, Chen X, Jia Q, Sheng Z, Wu A, Li W (2014) Proposal for fabrication-tolerant SOI polarization splitter-rotator based on cascaded MMI couplers and an assisted bi-level taper. Opt Express 22:27869–27879
42. Soldano LB, Pennings EC (1995) Optical multi-mode interference devices based on self-imaging: principles and applications. J Lightwave Technol 13:615–627
43. Kawaguchi Y, Tsutsumi K (2002) Mode multiplexing and demultiplexing devices using multimode interference couplers. Electron Lett 38:1701–1702
44. Ye M, Yu Y, Zou J, Yang W, Zhang X (2014) On-chip multiplexing conversion between wavelength division multiplexing-polarization division multiplexing and wavelength division multiplexing-mode division multiplexing. Opt Lett 39:758–761
45. Uematsu T, Ishizaka Y, Kawaguchi Y, Saitoh K, Koshiba M (2012) Design of a compact two-mode multi/demultiplexer consisting of multimode interference waveguides and a wavelength-insensitive phase shifter for mode-division multiplexing transmission. J Lightwave Technol 30:2421–2426
46. Maese-Novo A, Halir R, Romero-García S, Pérez-Galacho D, Zavargo-Peche L, Ortega-Moñux A, Molina-Fernández I, Wangüemert-Pérez JG, Cheben P (2013) Wavelength independent multimode interference coupler. Opt Express 21:7033–7040
47. Qiu C, Sheng Z, Li H, Liu W, Li L, Pang A, Wu A, Wang X, Zou S, Gan F (2014) Fabrication, characterization and loss analysis of silicon nanowaveguides. J Lightwave Technol 32:2303–2307
48. Sheng Z, Wang Z, Qiu C, Li L, Pang A, Wu A, Wang X, Zou S, Gan F (2012) A compact and low-loss MMI coupler fabricated with CMOS technology. IEEE Photonics J 4:2272–2277
49. Lin PT, Singh V, Wang J, Lin H, Hu J, Richardson K, Musgraves JD, Luzinov I, Hensley J, Kimerling LC, Agarwal A (2013) Si-CMOS compatible materials and devices for mid-IR microphotonics. Opt Mater Express 3:1474–1487
50. Nedeljkovic M, Khokhar AZ, Hu Y, Chen X, Penades JS, Stankovic S, Chong HMH, Thomson DJ, Gardes FY, Reed GT, Mashanovich GZ (2013) Silicon photonic devices and platforms for the mid-infrared. Opt Mater Express 3:1205–1214
51. Shankar R, Lončar M (2014) Silicon photonic devices for mid-infrared applications. Nanophotonics 3:4–5
52. Roelkens G, Dave UD, Gassenq A, Hattasan N, Hu C, Kuyken B, Leo F, Malik A, Muneeb M, Ryckeboer E, Sanchez D (2014) Silicon-based photonic integration beyond the telecommunication wavelength range. IEEE J Sel Top Quantum Electron 20:394–404
53. Milosevic MM, Nedeljkovic M, Ben Masaud TM, Jaberansary E, Chong HM, Emerson NG, Reed GT, Mashanovich GZ (2012) Silicon waveguides and devices for the mid-infrared. Appl Phys Lett 101:121105
54. Xia Y, Qiu C, Zhang X, Gao W, Shu J, Xu Q (2013) Suspended Si ring resonator for mid-IR application. Opt Lett 38:1122–1124
55. Van Camp MA, Assefa S, Gill DM, Barwicz T, Shank SM, Rice PM, Topuria T, Green WM (2012) Demonstration of electrooptic modulation at 2165 nm using a silicon Mach-Zehnder interferometer. Opt Express 20:28009–28016
56. Nedeljkovic M, Stankovic S, Mitchell CJ, Khokhar AZ, Reynolds SA, Thomson DJ, Gardes FY, Littlejohns CG, Reed GT, Mashanovich GZ (2014). Mid-infrared thermo-optic modulators in SOI. IEEE Photonics Technol Lett 26:1352–1355
57. Souhan B, Grote RR, Chen CP, Huang HC, Driscoll JB, Lu M, Stein A, Bakhru H, Bergman K, Green WM, Osgood RM (2014) Si^+-implanted Si-wire waveguide photodetectors for the mid-infrared. Opt Express 22:27415–27424
58. Muneeb M, Chen X, Verheyen P, Lepage G, Pathak S, Ryckeboer E, Malik A, Kuyken B, Nedeljkovic M, Van Campenhout J, Mashanovich GZ (2013) Demonstration of silicon-on-insulator mid-infrared spectrometers operating at 3.8 μm. Opt Express 21:11659–11669
59. Hu Y, Li T, Thomson DJ, Chen X, Penades JS, Khokhar AZ, Mitchell CJ, Reed GT, Mashanovich GZ (2014) Mid-infrared wavelength division (de)multiplexer using an interleaved angled multimode interferometer on the silicon-on-insulator platform. Opt Lett 39:1406–1409

60. Cheng Z, Chen X, Wong CY, Xu K, Fung CK, Chen YM, Tsang HK (2012) Focusing subwavelength grating coupler for mid-infrared suspended membrane waveguide. Opt Lett 37:1217–1219
61. Lin PT, Singh V, Lin HYG, Tiwald T, Kimerling LC, Agarwal AM (2013) Low-stress silicon nitride platform for mid-infrared broadband and monolithically integrated microphotonics. Adv Opt Mater 1:732–739
62. Tai Lin P, Singh V, Kimerling L, Murthy Agarwal A (2013) Low-stress silicon nitride platform for mid-infrared broadband and monolithically integrated microphotonics. Appl Phys Lett 102:251121
63. Soref R (2010) Mid-infrared photonics in silicon and germanium. Nat Photonics 4:495–497
64. Sacher WD, Huang Y, Ding L, Barwicz T, Mikkelsen JC, Taylor BJ, Lo GQ, Poon JK (2014) Polarization rotator-splitters and controllers in a Si$_{34}$ 4-on-SOI integrated photonics platform. Opt Express 22:11167–11174
65. Kischkat J, Peters S, Gruska B, Semtsiv M, Chashnikova M, Klinkmüller M, Fedosenko O, Machulik S, Aleksandrova A, Monastyrskyi G, Flores Y (2012) Mid-infrared optical properties of thin films of aluminum oxide, titanium dioxide, silicon dioxide, aluminum nitride, and silicon nitride. Appl Opt 51:6789–6798
66. Wang J, Lee C, Niu B, Huang H, Li Y, Li M, Chen X, Sheng Z, Wu A, Li W, Wang X (2015) A silicon-on-insulator polarization diversity scheme in the mid-infrared. Opt Express 23:15029–15037

Chapter 5
CMOS-Compatible Efficient Fiber-to-Chip Coupling

5.1 Research Progress of Si Inverse Taper

Although various passive or active devices have been realized in SOI waveguides, the coupling problem between optical fiber and SOI waveguide must be solved in addition to the silicon-based optical interconnections. In the above chapter, we discuss the polarization problem between fiber and waveguide. In this chapter, we further discuss the coupling loss problem.

There are many sources of coupling loss between optical fiber and waveguide. In addition to the alignment error caused by the position or angle offset, the differences in the two modes also cause large loss. Figure 5.1 shows the schematics of cross-sections of the typical silicon nanowire optical waveguides and standard single-mode optical fibers. The mode area of in the waveguide is generally less than 1 μm, while the mode field diameter (MFD) of the standard single mode fiber are about 10.4 μm and 9.2 μm at the wavelength of 1550 nm and 1310 nm respectively. Moreover, the effective refractive index difference between the two modes is also very large. Therefore, when light enters the waveguide from the optical fiber, the mismatching of the mode size causes the radiation mode and some scattering loss. The difference of the effective refractive index causes the reflection and the some reflection loss. Therefore, the coupling losses between optical fiber and waveguide often reach 20 dB.

At present, there are three coupling schemes for optical fibers and waveguides in silicon photonics. The first one is called edge coupling, or in-plane coupling. The second one is called the vertical coupling, also called grating coupling or the out-plane coupling. The third one is called the prism coupling [1, 2]. The edge coupling and grating coupling are widely studied by the researchers. The content of grating coupling and prim coupling is not detailed in this article due to the chapter length. In this section, we will review the latest progress of the edge coupling technology based on inverse taper.

J. Wang, *CMOS-Compatible Key Engineering Devices for High-Speed Silicon-Based Optical Interconnections*, Springer Theses,
https://doi.org/10.1007/978-981-13-3378-1_5

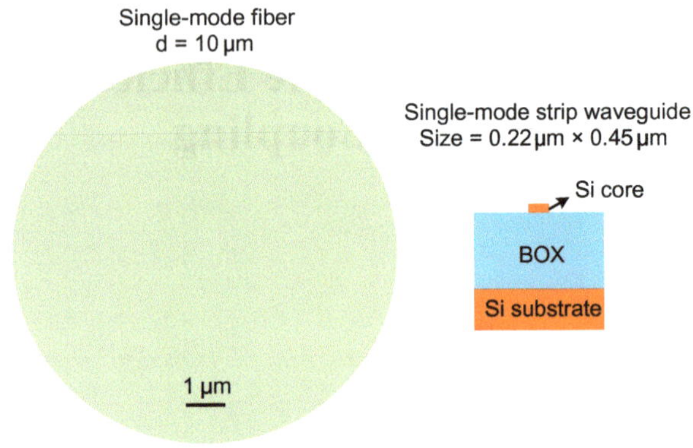

Fig. 5.1 Comparison of standard single-mode fiber and silicon nanowire optical waveguide

Inverse taper is a kind of edge coupling technology. For normal single-mode SOI waveguides, due to the large refractive index difference between the core Si and the cladding SiO_2, the mode field is mainly confined to the Si core layer. The limiting effect on light will weaken as the width or height of the waveguide is reduced. Especially when the waveguide size is only a few dozens of nm, a large number of mode field is leaked into the cladding. On the one hand, the mode size is expanded, which is closer to the MFD of the optical fiber. On the other hand, the effective refractive index of the mode is also reduced as more field enters the cladding. It makes it closer to the effective refractive index of the optical fiber mode, which is beneficial to reducing the loss of reflection loss.

The initial inverse coupler is relatively simple. Only the width of the Si waveguide is narrowed, and its cladding structure is not changed. Or while the width of the waveguide is reduced, other low-refractive-index materials are used to form a new waveguide. In this structure, the new waveguide has only one layer. In order to distinguish from the latter bi-level structure, I call this simple inverse taper single-layer one.

In 2003, Almeida et al. from Cornell university realized a inverse taper based on silicon nanowire optical waveguide [3]. When the tip width is about 100 nm, the length is 40 μm and the wavelength is 1.55 μm, the coupler loss for TE and TM are 6 dB and 3.3 dB, respectively. In 2014, Cardenas and others in the same group improved the process, and studied the effect of the gap between the tip and the chip edge on the coupling efficiency, and finally got low-loss of 0.7 dB per facet [4]. The structure of the inverse taper in this research group is very simple. Only need to change the width of the waveguide and the upper cladding is still SiO_2, so it is a inverse taper that is truly compatible with the CMOS process. In 2002, Shoji and others from NTT used a 3 μm × 3 μm polymer square waveguide outside the inverse taper, so that the waveguide mode was closer to the optical fiber mode and the coupling loss was only 0.8 dB [5]. In 2005, Tsuchizawa et. al., in NTT used SiON and SiO_2as the core and cladding of a new waveguide respectively when the width of the Si waveguide was reduced [6]. The test results show that for MFD = 4.3 μm and 9 μm, the coupling loss are only 0.5 dB and 2.5 dB respectively. In 2003,

Mcnab et al., from IBM used a 2 μm×2 μm polymer square waveguide outside the inverse taper and the coupling loss is less than 1 dB [7]. In 2009, Wahlbrink et al., used a 5 μm×3 μm SU-8 square waveguide outside the inverse taper. The maximum coupling efficiency is over 60 %, and 3 dB bandwidth exceeds 100 nm [8]. In 2010, Pu et al., from DTU used thermal oxidation technology to reduce the width of the taper tip to less than 15 nm, and covered it by 3.4 μm thick SU-8 waveguide. The test results show that for TE and TM modes, the coupling loss is only 0.66 dB and 0.36 dB repectively [9]. In 2011, Ren et al., studied the influence of the contour shape of the inverse taper on the performance. It was found that the structure with exponential or square linear shape is superior to the linear inverse taper on coupling efficiency, footprint, process tolerance and misalignment tolerance [10].

However, the above single-layer inverse taper is generally suitable to couple with the lensed fiber with small MFD, and the coupling loss with the standard single-mode fiber will be very large. In order to further extend the mode field to match with single-mode fibers, two-layer or even three-layer waveguides can be constructed. Such structures can not only be realized in the linear taper whose core size is kept constant or increased, and can also be used in the inverse taper [11–15].

In 2010, Khilo et al., from MIT proposed to use low-refractive-index material as a bi-level stage, which is connected to a Si inverse taper [16]. The simulation results show that the coupling loss is only 0.5 dB for MFD = 8 μm. The design was demonstrated by Park et al., from ETRI in 2013 [17]. The test results show that for TE and TM modes, the average coupling loss are 4.2 dB and 3.6 dB respectively for the single-mode fiber with MFD = 10 μm. Although the measured loss is large, it is still expected the loss can be further reduced by improving the process. In 2012, Tokushima et al., from AIST realized a rib inverse taper by using SiO_2 as the core layer in the low-refractive-index rib waveguide [18]. Si layer is also reduced in the vertical direction. The test results show that for TE and TM modes, the coupling loss is 2.7 dB and 3 dB for single-mode fibers with MFD = 10 μm. In 2012, Shiraishi et al., realized a inverse taper with size reducing in the vertical direction by using dry etching process [19]. The test results show that the coupling loss for TE and TM modes is 2.8 dB and 2.7 dB for single-mode fibers with MFD = 10 μm. In 2013, Ku et al., used a Si inverse taper and a three dimensional SU-8 inverse taper to achieve an efficient fiber-to-coupler [20]. The test results show that for TE and TM mode, the coupling loss for TE and TM modes is 2.8 dB and 4.1 dB for single-mode fibers with MFD = 10 μm.

As the single mode fiber has a large MFD, some mode field will leak to the Si substrate. In addition, the infinite width in the buried oxide layer also makes the waveguide mode field at the edge of the inverse taper not symmetrical, causing a large difference from the optical mode field of the optical fiber. Therefore, the researchers proposed that by etching part of buried oxide layer and Si substrate, the cantilever structure can be achieved, so that the waveguide mode is closer to the optical fiber mode. Moreover, etching deep groove will help the packaging of the optical fiber.

In 2009, Sun et al., used reactive ion etching to achieve a suspended Si inverse taper with a cladding of SiO_2 [21]. The test results show that for TE and TM modes, the coupling loss are 1.6 dB and 2 dB respectively. In 2012, Wood et al., from the same group further optimized the device design to achieve a suspended Si inverse taper with a length of only 6.5 μm [22] . The test results show that in the C band, the coupling loss are 0.62 dB and 0.5 dB for TE and TM modes respectively. In 2010, Bakir et al., also used the Si-rich SiO_x to realize a similar Si suspended inverse taper [23]. The test results show that the coupling loss with lensed fiber is less than 1 dB in the wavelength range from 1520 to 1600 nm. In 2010, Chen et al., filled a kind of low-refractive-index oil-based material around a suspended Si inverse taper and added a number of suspended structures [24]. The test results show that the coupling loss with the single-mode fiber is about 1.5 dB–2 dB. In 2010, Fang et al., adopted a bi-layer SiO_2 waveguide with bi-level Si inverse taper to realize the suspended fiber-to-chip inverse taper [25]. The test results show that the coupling loss with the single mode fiber is less than 4 dB. In 2011, they further increased the thickness of the top Si and SiO_2 to achieve a better mode matching with the single-mode fiber mode [26]. The test results show that the coupling loss of TE and TM modes are 1.2 dB and 2 dB for single-mode optical fibers, respectively, in case of refractive-index matching liquids. In 2014, Jia et al., used SiON as a transitional material to alleviate the requirement for Si taper tip width [27]. The test results show that for TE and TM modes, the coupling loss with single-mode fiber are 1.8 dB and 2.1 dB respectively.

Whether it is a double-layer, three-dimensional or suspended structure, it is mainly aimed at improving the low-refractive-index waveguides. Some researchers also put

Table 5.1 Summary of the experimental results of Si inverse taper(2003–2015.5)

Group and time	Loss[a]	Tolerance[b]
Almeida [3]/Cornell/2003	3.3/6.0/5	±1.2/±1.2/1/TE
Cardenas [4]/Cornell/2014	0.7/*/2.5	*
Shoji [5]/NTT/2002	0.8/*/4.3	*
Tsuchizawa [6]/NTT/2005	0.5/*/4.3	*
Mcnab [7]/IBM/2003	1/2/2.1	±1/±1/3/TE
Wahlbrink [8]/AMO/2009	2.51/2.21/2.5	*
Pu [9]/DTU/2010	0.66/0.36/2.9	±1.5/±1.5/3/TM
Park [17]/ETRI/2013	4.2/3.6/10	±2.2/±2.2/1/TE
Tokushima [18]/AIST/2012	2.7/3.0/10	±1.2/±1.2/0.25/TM
Shiraishi [19]/Utsu. Univ./2012	2.8/2.7/5.2	*
Ku [20]/NTHU/2013	2.8/4.1/10	±3/±3/3/TE
Sun [21]/OSU/2009	1.6/2/1.5	*
Wood [22]/OSU/2012	0.48/0.39/1.5	*
Bakir [23]/CEA-LETI/2010	0.25/0.25/3	*
Fang [25]/ASTAR/2010	1.85/2.2/10.5	±1.7/±1.7/1/TE
Fang [26]/ASTAR/2011	1.2/2.0/10.5	±2.8/±2.1/1/TE
Jia [27]/ASTAR/2014	1.8/2.1/9.2	±3.5/±3/3/TE

[a]TE loss (dB)/TM loss (dB)/MFD (μm)
[b]Horizontal tolerance (μm)/Vertical tolerance (μm)/Bandwidth (dB)/Polarization
*denotes no performance characterization from the reference in the table

forward a series of novel design schemes about the Si tapered waveguide itself, such as double-tip waveguide [28, 29], slot waveguide [30, 31] and sub-wavelength grating waveguide [32]. Table 5.1 summarizes the experimental results of the reported Si inverse taper.

5.2 Fiber-to-Chip Coupler Based on Single-Tip Inverse Taper

According to the summary of the research progress of inverse taper in the last section, the structure of the inverse taper is relatively complicated. The structure of multi-stage, multi-layer and cantilever makes it difficult to realize under the standard CMOS process. The device reported by Cornell university is a structure that can really be processed under the CMOS platform, as shown in Fig. 5.2. Our inverse taper also uses a similar structure.

We designed the device on the SOI wafer with a top silicon thickness of 220 nm and a buried oxide layer of 2 μm. In order to ensure the symmetry of the waveguide mode, the upper cladding also uses approximately 2 μm thick SiO_2. The inverse taper has a tip width of W_{tp} and a length of L_{tp}. The width is linearly increased to 0.45 μm at the end of the coupler.

For an inverse taper, it is very important to obtain a smooth high-quality taper tip. Mechanical polishing can only deal with one chip at a time, and the efficiency is very low. The traditional etching technology will simultaneously etch many kinds of materials, such as SiO_2 upper cladding and Si waveguide. Different etching rates

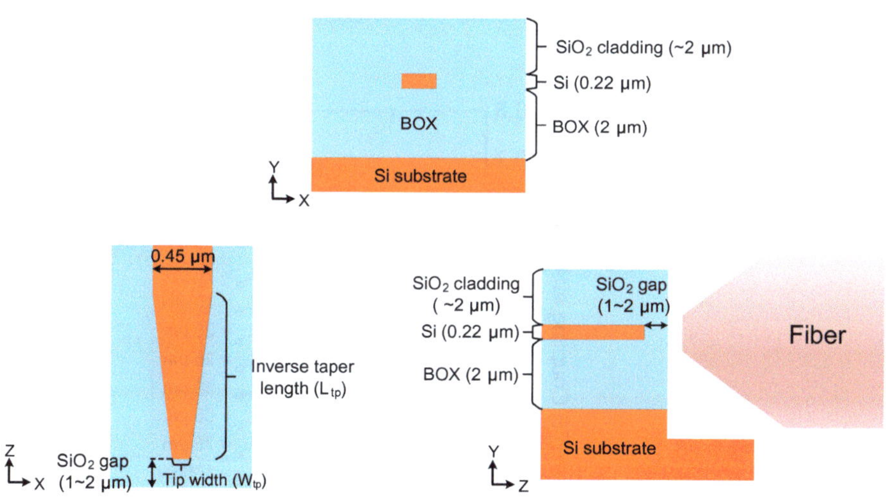

Fig. 5.2 Schematic of a single-tip inverse taper

will destroy the quality of the taper tip, resulting in coupling loss. In order to obtain a high-quality fiber-to-waveguide interface, we leave 1 to 2 μm SiO_2 gap at the Si taper tip and the chip edge, thus ensuring that only the cladding material will be etched and the vertical Si cone tip is obtained. Moreover, this technology can be processed on the whole wafer, and the efficiency is very high.

The coupling loss between inverse taper and optical fiber mainly comes from three parts, which are the mode mismatching loss between the waveguide and optical fiber, the mode conversion loss in inverse taper and the propagation loss caused by the surface roughness of the waveguide. In the simulation of waveguide transmission loss, a more complicated waveguide roughness model is needed. The related work can be seen in the literature [33]. Here, we only consider the first two kinds of losses in the simulation.

Figures 5.3 and 5.4 show the relationship between the TE and TM coupling losses and the tip width W_{tp} with different coupler length L_{tp} respectively when the wavelength is 1550 nm. Here, we used a three-dimensional FDTD to calculate the mode conversion loss from a inverse taper to a single mode waveguide with a width of 0.45 μm (i.e., the coupling loss) for the Gauss beam of MFD = 3.3 μm.

For the taper tip width W_{tp} with small value (such as 40 nm) or large value (160 nm), the relationship between the coupling efficiency and the device length $L_{tp}$is the opposite. Take the TE mode as an example.

- When W_{tp} = 40 nm, the shorter the device L_{tp}, the better the performance. This is because the small waveguide width makes the waveguide mode like the slab mode at this time, so the optical fiber mode is mismatched with the waveguide mode, and a considerable part of the power is radiated out. In order to coupling with these radiated power, the width of the coupler needs to be increased quickly for better mode matching, so the shorter the length of the coupler, the better the performance of the coupler.

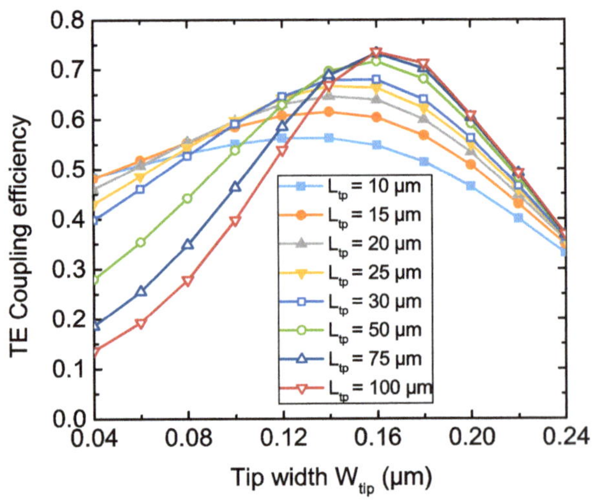

Fig. 5.3 Simulated TE loss of a inverse taper

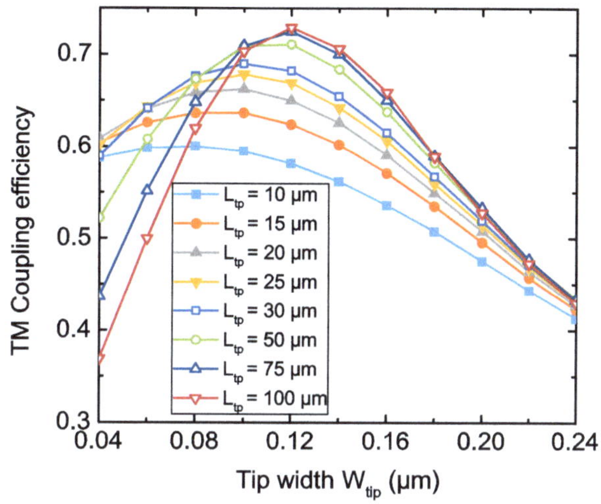

Fig. 5.4 Simulated TM loss of a inverse taper

- When $W_{tp} = 160$ nm, the shorter the device L_{tp}, the worse the performance. Because the mode mismatching loss is small, a small amount of radiation power will soon be tilted to other directions. The coupling loss is mainly determined by the conversion loss in the coupler, and the longer coupler has the smaller mode conversion loss. Moreover, it can be found that the coupling efficiency will slowly converge as the $L_{tp}$of the coupler length increases.

For the TE mode, the best coupling efficiency can be obtained when the taper tip width is $W_{tp} = 160$ nm; for the TM mode, the optimal value is obtained at the taper tip width of $W_{tp} = 120$ nm.

These inverse tapers are fabricated on the SOI wafer with 220 nm top silicon layer and 2 μm buried oxide layer. The waveguide structure is defined by electron beam exposure, and a 2 μm thick SiO_2 layer is deposited. Finally, the edge of the chip is processed by reactive ion deep etching.

In order to extract the coupling loss, we fabricated some single mode waveguides with 0.45 μm ×0.22 μm cross-section size and different lengths. These waveguides have the same bend and inverse tapers, only the length of the straight waveguide is different. We tested the transmission spectrum of these waveguides at wavelength of 1550 nm and performed a linear fitting. The propagation loss of the waveguide are 2.31 dB/cm and 1.35 dB/cm for TE and TM modes, respectively (Fig. 5.5).

In order to make full use of the layout area, we put the inverse taper layout as shown in Fig. 5.6a. A group of devices with the same W_{tp} and L_{tp} design parameters have 4 devices to prevent process failure. Moreover, each device includes a curved waveguide with a radius of 50 μm to prevent input light from entering the optical fiber directly. The gap between adjacent devices is 50 μm to remove unnecessary coupling. Figure 5.6b on the layout shows the actual enlarged picture of the processed inverse taper.

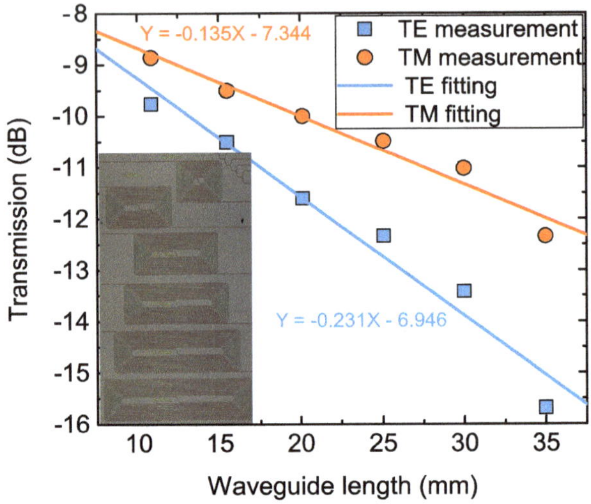

Fig. 5.5 Transmission measurement results of waveguides with different lengths at 1550 nm wavelength

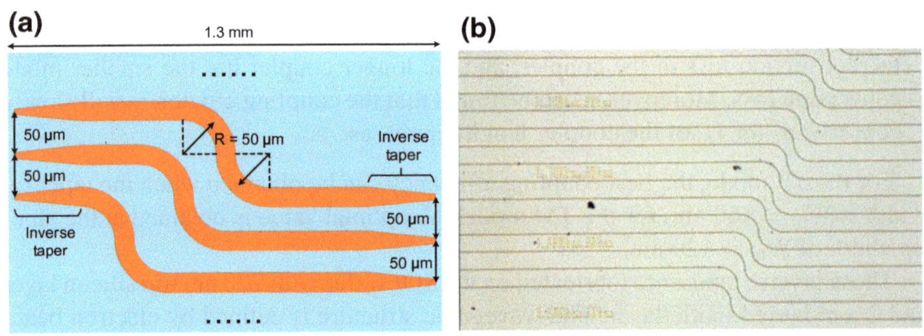

Fig. 5.6 **a** Schematic of the test structure of the inverse taper on the layout. **b** Microscoped picture of the single-tip inverse taper sample

The coupling loss is extracted as the following steps: First, test the fiber-to-fiber loss L_{f2f} without chip. Then, test the total loss L_{total} in the whole link. Finally, according to the propagation loss previously obtained, the loss L_s of the straight waveguide and bend waveguide on the chip is calculated. Because the radius of the bend waveguide is large, it can be treated as a straight waveguide. In this way, the coupling loss is (L_{total}-L_{f2f}-L_s) /2 (dB/facet). In addition, a more accurate method is to design a series of links with straight waveguides in different lengths, so that the link loss without straight waveguide can be obtained by linear fitting, that is, only the coupling loss and fiber-to-fiber loss. Then the fiber-to-fiber loss can be removed to obtain the coupling loss. However, the second method requires more testing structures.

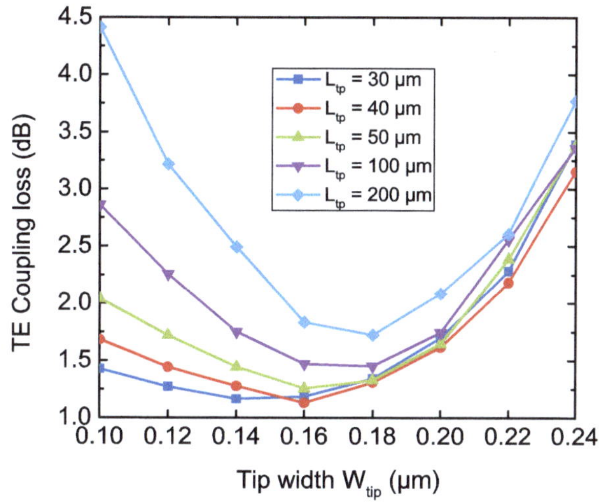

Fig. 5.7 Measured TE loss of a inverse taper

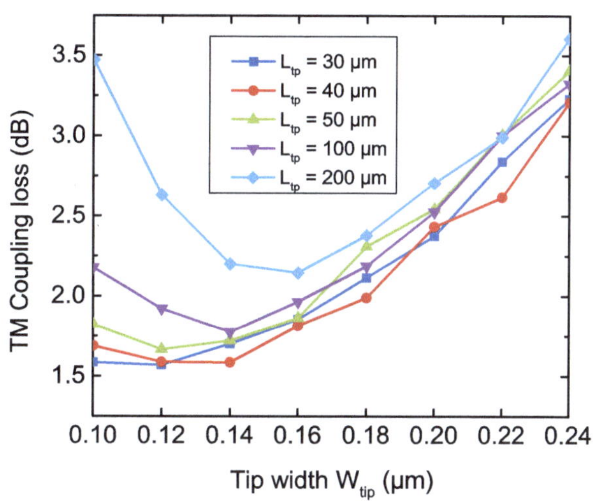

Fig. 5.8 Measured TM loss of a inverse taper

Figures 5.7 and 5.8 show the TE and TM coupling loss when the wavelength is 1550 nm respectively. When $W_{tp} = 160$ nm and 120 nm, the TE and TM coupling loss are the smallest, respectively. This is very close to our simulation results. However, the length L_{tp} with minimum coupling loss is very different from that in simulation. For TE and TM modes, when L_{tp} is 40 μm and 30 μm, the minimum coupling loss can be 1.13 dB/facet and 1.57 dB/facet, respectively. Moreover, the coupling loss increases with L_{tp}. This shows that the propagation loss of the coupler has great effect on coupling loss. The Si core cann't confine light well because the waveguide width is small. A lot of mode field are distributed on the waveguide surface, which will cause a lot of scattering loss. The scattering loss is even higher than the mode conversion loss in practical devices.

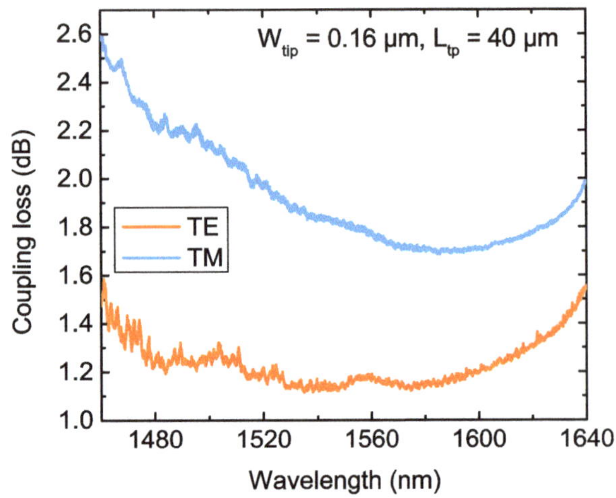

Fig. 5.9 Wavelength dependence of the coupling loss of the inverse taper when the taper tip width $W_{tp} = 0.16\ \mu m$ and the length $L_{tp} = 40\ \mu m$

Finally, we tested the wavelength dependence of the coupling loss of the inverse taper width tip width $W_{tp} = 0.16\ \mu m$ and the length $L_{tp} = 40\ \mu m$, as shown in Fig. 5.9. It is found that the coupling loss exhibits large broadband characteristics. For the TE mode, the coupling loss is only 0.5 dB in the range of 1460 to 1640 nm. For the TM mode, there is less than 1 dB fluctuation.

5.3 Ultra-Compact Optical Splitter Based on Two-Mode Interference (TMI)

Optical splitter or combiner is one of the most important devices in integrated optics. In silicon photonics, there have been some recent important research developments recently. In 2012, Sheng et al., realized an ultra-low loss MMI splitter fabricated with 0.13 μm CMOS process [34]. For the TE mode, the excess loss at a wavelength of 1550 nm is only 0.06 dB. In 2013, Zhang et al. used particle swarm optimization to demonstrate a novel Y-junction optical splitter. The size of the device is only 1.2 μm ×2 μm, and the excess loss is only 0.28 dB [35]. In 2013, Xiao et al., realized the a s 1.5 μm × 1.8 μm splitter by using the TMI effect [36]. For TE and TM modes, the excess loss are only 0.11 dB and 0.18 dB respectively.

In all kinds of optical splitter structures, MMI coupler has relatively large bandwidth and high processing tolerance. In this section, we hope to design a polarization-insensitive and low-loss optical splitter. Therefore, we also refer to the design of Sheng and Xiao.

Figure 5.10 shows the schematic of MMI design. Its multi-mode waveguide width is W_{MMI} and its length is L_{MMI}. The width of the input and output waveguides increases linearly from W_{co} to W_t, and the length is L_{MMI_tp}. The gap between the two output waveguides near the multi-mode waveguide region is W_g. In order to realize

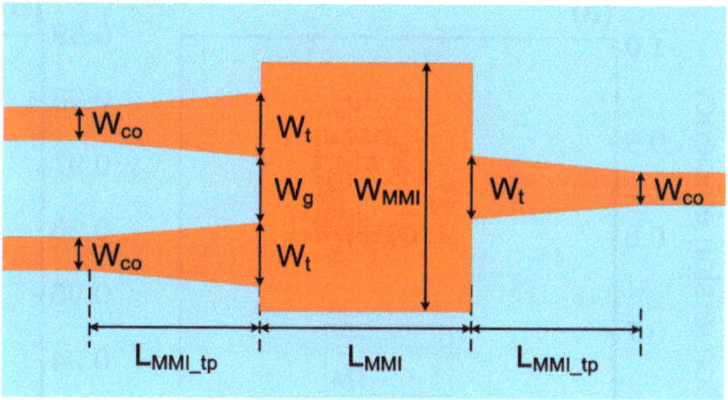

Fig. 5.10 Schematic of the MMI design

the polarization-insensitive MMI splitter, in the case of symmetric interference, the width of the multimode region should only support the lowest two order modes, that is, TE_0/TE_2, TM_0/TM_2. More theoretical analysis can refer to Xiao's thesis. Here, we choose $W_{MMI} = 1.6$ μm to ensure the waveguide width can satisfy. In order to reduce the mode conversion loss in the multi-mode region, W_t should be large enough. Considering the limitation of the process precision, $W_t = 0.75$ μm is used here, so that the corresponding $W_g = 0.05$ μm can be processed in the case of electron beam lithography. However, it must be pointed out that for the standard CMOS process, the requirement of W_g is more stringent. So designing such a splitter needs more consideration and optimization.

We use FIMMPROP to optimize other parameters of MMI. Figure 5.10a shows the conversion efficiency for two input polarization. It can be found that the optimal TE mode conversion efficiency can be obtained when $L_{MMI\_tp} = 2.1$ μm. Although the optimal L_{MMI_tp} for TM conversion efficiency will shift to about 1.9 μm, the TM conversion efficiency is still 94% at $L_{MMI\_tp} = 2.1$ μm. So here we choose $L_{MMI\_tp} = 2.1$ μm. In order to reduce the mode conversion loss in the input and output waveguides, L_{MMI_tp}also needs to be optimized, otherwise it is possible that this loss may be even higher than the mode conversion loss in the multi-mode region. Figure 5.10b shows the relationship between conversion efficiency and L_{MMI_tp} for TE mode. Here, we choose $L_{MMI\_tp} = 1.5$ μm to get the conversion efficiency of about 98.5%. Table 5.2 summarizes the design parameters of this MMI splitter. The total length of the whole MMI is only 5.1 μm (Fig. 5.11).

These devices are processed with the inverse taper in the previous section under the same process. In order to test the excess loss of the MMI splitter, it can be cascaded by several MMI modes, as shown in Fig. 5.12. We have cascaded 6 MMI couplers

Table 5.2 Design parameters of MMI

W_{co} (μm)	L_{MMI_tp} (μm)	W_t (μm)	W_g (μm)	W_{MMI} (μm)	L_{MMI} (μm)
0.45	1.5	0.75	0.05	1.6	2.1

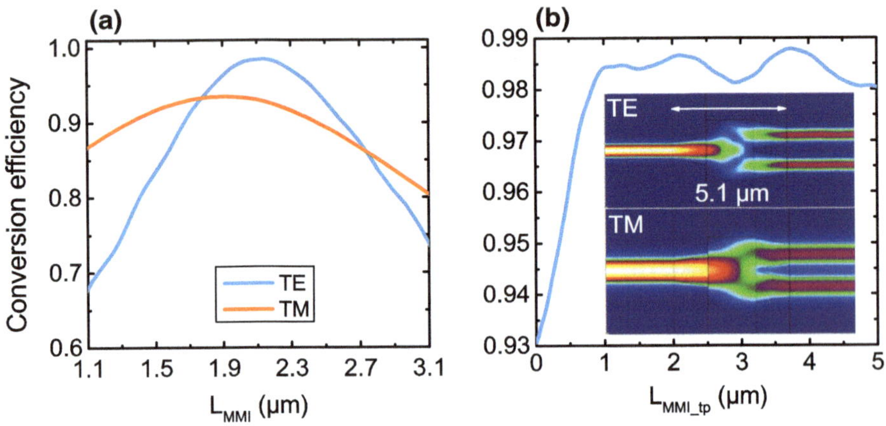

Fig. 5.11 Simulation result of the MMI splitter at 1550 nm wavelength

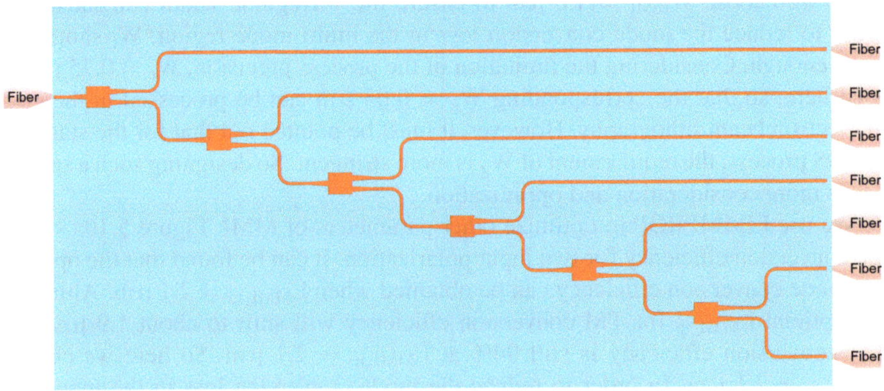

Fig. 5.12 Schematic of the cascaded structure to test the excess loss of the MMI splitter

and the excess loss can be accurately obtained by testing the transmission spectrum in each output port.

Figure 5.13 shows the test results of the cascaded MMI splitter in each port at wavelength of 1550 nm. We use edge coupling to test these MMI couplers, and the inverse taper on each link has the same design parameters. These data points are well fitted linearly and show that the test results are reliable. Moreover, for different input polarizations, the excess loss is less than 0.1 dB, which proves that the polarization-insensitive design is correct.

We also tested the transmission spectrum of the cascaded MMI splitter, as shown in Fig. 5.14. These test results also show wavelength dependence of the other parts of the whole link. Because the inverse taper and waveguide are broadband, it can also reflect the MMI has a relatively large bandwidth.

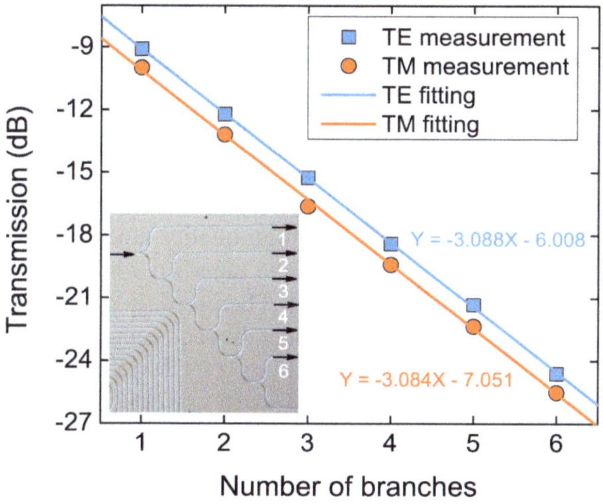

Fig. 5.13 Measured results of cascaded MMI splitter at 1550 nm wavelength

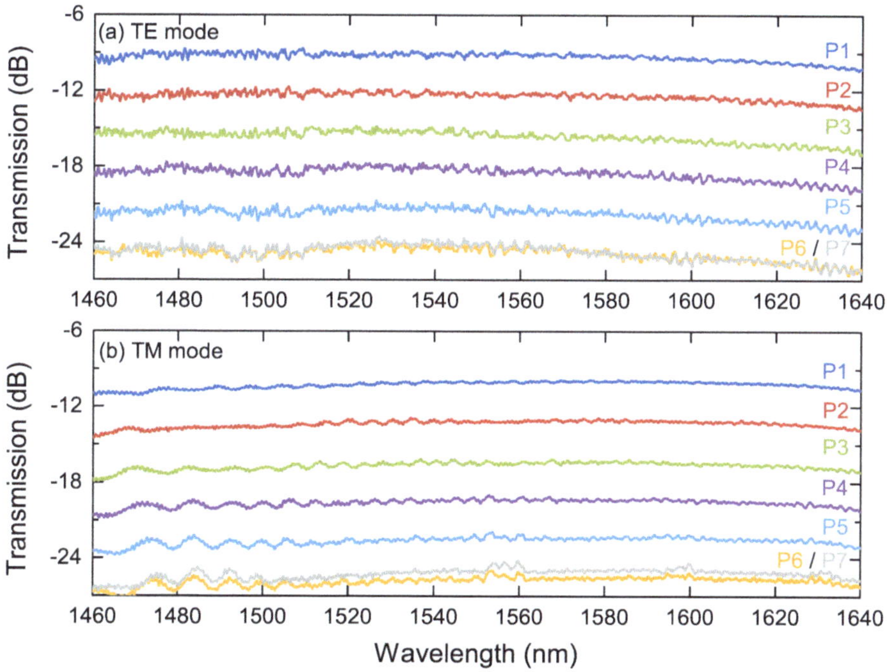

Fig. 5.14 Wavelength dependence of the performances of the cascaded MMI splitter

5.4 Double-Tip Edge Coupler Based on Polarization-Insensitive TMI Combiner

Up to now, most of the inverse tapers have only one taper tip. For the inverse taper with a tip width of only a few dozen nm or about 100 nm, the waveguide roughness and the difference between the actual truncation position and the designed position will bring great difficulty to realizing a high-quality taper tip. In order to obtain smooth facet, it is very likely to damage the taper tip when mechanical polishing or etching is done on the waveguide facet. Therefore, in order to realize the high-performance edge coupling scheme and meet the requirements of low coupling loss, large bandwidth, small size and compatibility of CMOS process simultaneously, it needs to be further optimized. One way is to use double-tip inverse taper. One of its prominent advantages is that when a taper tip is damaged, another good tip can still maintain a reasonable coupling loss. Moreover, the double-tip inverse taper provides a new design freedom, such as the waveguide gap of the taper tip, which is used to improve the device performance. The double-tip inverse taper has been experimentally demonstrated in the SiN platform, but its coupling loss is 2 dB per facet, and the length is up to 400 μm [28]. The edge coupler design of the Y-junction on the SOI platform was also proposed, but there was no prototype device [29].

In this section, we use two inverse tapers and a polarization-insensitive TMI optical combiner to demonstrate a double-tip fiber-to-chip edge coupler for the first time on the SOI platform. We have selectively processed some inverse tapers with different lengths, tip widths and taper tip gap, and characterized them. The test results show that for a coupler with a length of only 40 μm, the coupling loss per facet are only 1.10 dB and 1.52 dB respectively for TE and TM modes.

Figure 5.15 shows the schematic of our proposed double-tip fiber-to-chip coupler. The whole device is designed on the SOI wafer with 220 nm top silicon layer and SiO_2 upper and lower cladding with 2 μm thickness. It consists of two inverse tapers and a TMI optical combiner. Two inverse tapers form a double-tip inverse taper, which has a taper tip width of W near the fiber facet. The waveguide width will gradually increase from W to 0.45 μm, while the tip gap is G. The two waveguides will be combined into one and then connected with other integrated devices through a TMI optical combiner. One of the difficulties of designing a double-tip inverse taper is a

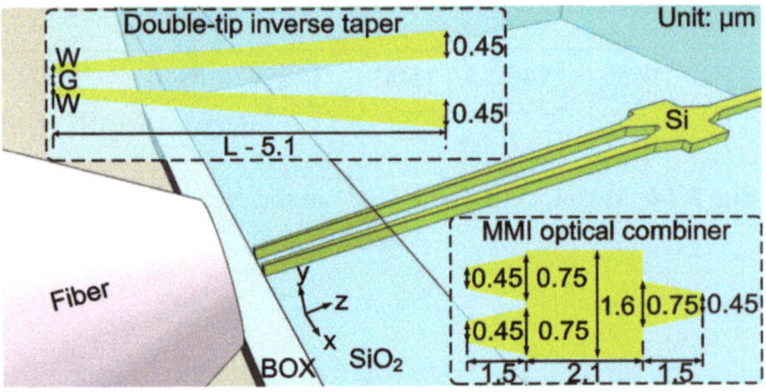

Fig. 5.15 Schematic of the fiber-to-chip edge coupler based on double-tip inverse taper and polarization-insensitive TMI optical combiner. Reproduced from Ref. [37] by permission of Optical Society of America

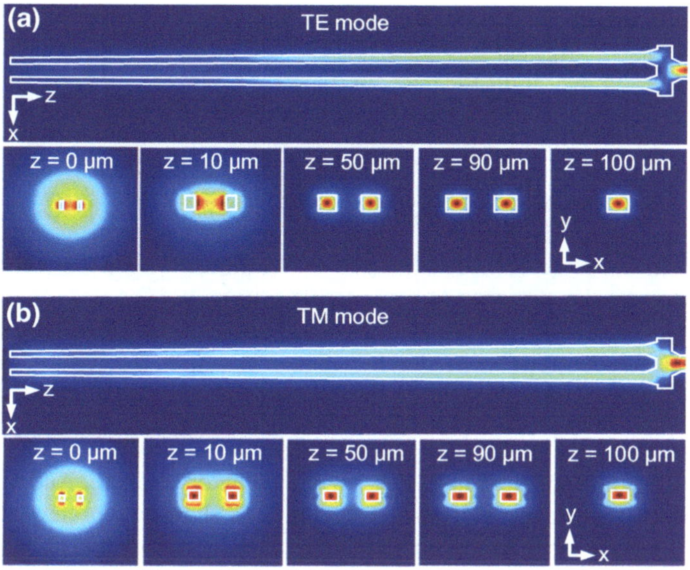

Fig. 5.16 Mode propagation in a 100 μm-long double-tip inverse taper. [2016] IEEE. Reprinted, with permission, from Ref. [38]

optical combiner with low excess loss and low polarization-dependent loss. Here we use the TMI optical combiner mentioned in the previous section. The optimization parameters are shown in the internal diagram of Fig. 5.15. It can satisfy our design requirements. The length is only 5.1 μm and the excess loss for two polarizations is less than 0.1 dB.

We use three-dimensional FDTD to simulate the mode propagation in a inverse taper with a length of L = 100 μm, a taper tip width W = 140 nm, a taper tip gap G = 400 nm, as shown in Fig. 5.16. The simulated wavelength is 1550 nm and the light source has Gaussian beam with MFD = 3.3 μm, which is consistent with our actual lensed fiber. We also show the mode distribution in different directions along the propagation direction. As we can see, for the TE and TM modes, the light from the external optical fiber will first be coupled to the super-mode of the tip at the chip edge, which has an approximate circular mode distribution. Then, as the waveguide width and gap increase gradually, the mode distribution will also change, and more mode fields are confined in the waveguide core layer. Moreover, it can be found that the optimized TMI combiner works well and only a small amount of light is scattered out.

We fabricated these inverse taper samples on the SOI wafers with 220 nm top silicon and a 2 μm buried oxide. The test structure is similar to that in the second section, processing four identical devices for each design parameter to ensure the yield. The process steps are as follows. First, a layer of 100 nm HSQ negative photoresist is spined on the SOI wafer. Then the waveguide structure is defined

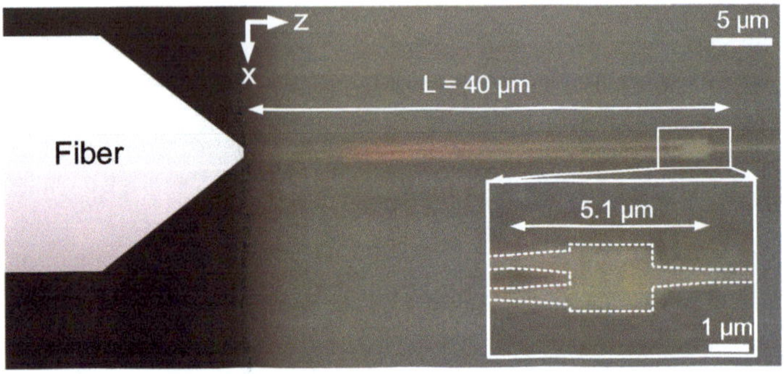

Fig. 5.17 Microscoped picture of the double-tip inverse taper with 40 μm length. Reproduced from Ref. [37] by permission of Optical Society of America

by electron beam exposure. The Si layer is etched by dry etching. And then a layer of 2 μm thick SiO_2 layer is deposited. Figure 5.17 shows a top-view microscope picture of a inverse taper with a length of 40 μm, W = 140 nm and G = 400 nm. In order to extract the coupling loss of the device, we also use the method introduced in second section to process a series of single-mode waveguides with different lengths, and to fit the propagation loss of the waveguide.

In addition to the design parameters of MMI coupler, there are three other parameters in the whole coupler, i.e., W, G and L, which have the greatest effect on device performances. Because in the second section, we find that due to the propagation loss of waveguide, the performance of the shorter coupler is better. Therefore, we chose a set of coupler with length L = 40 μm to study the effect of W and G on coupling loss. Figures 5.18 and 5.19 show the relationship between the TE and TM coupling loss and the taper tip width W and the gap G respectively when the wavelength is 1550 nm, length L = 40 μm and MFD = 3.3 μm. For the TE mode, when W = 120 nm ~160 nm, the minimum coupling loss can be obtained. And when G = 500 or 600 nm, the coupling loss is very stable in this W range, only 0.2 dB variation. Although the propagation loss of the waveguide is reduced with the increasing of the taper tip width W, the coupling loss is generally increasing because of the larger taper tip width resulting in better mode mismatching loss. For the TM mode, it is found that the coupling loss increases when W decreases or G increases. This is because the mode mismatching loss is larger at this time. We also give a simulation result of the coupling loss that does not include the propagation loss of the waveguide, which is in good agreement with the test results. The difference between them mainly comes from the propagation loss of the waveguide. We need to pay special attention to the fact that because of the new design freedom G in this coupler, it also has the advantage of low polarization-dependent loss. It can be found that when W = 140 nm and G = 600 nm, the polarization-dependent loss of the device is only 0.16 dB. Therefore, we believe that if we further optimize this device, it is entirely possible to obtain polarization-insensitive fiber-to-chip edge coupler.

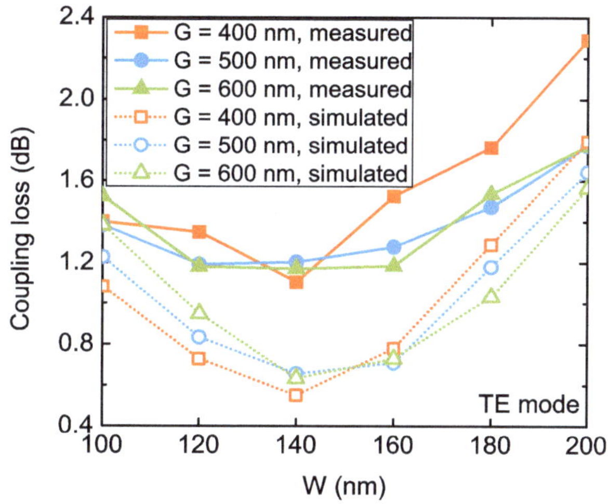

Fig. 5.18 Relationship between the TE coupling loss and taper tip width with different waveguide gap. Reproduced from Ref. [37] by permission of Optical Society of America

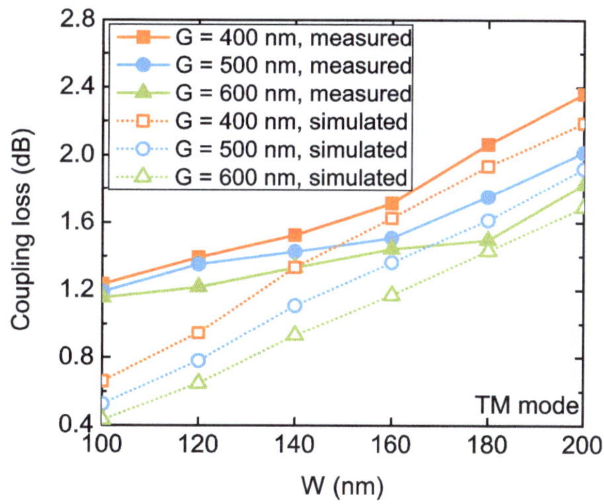

Fig. 5.19 Relationship between the TE coupling loss and taper tip width with different waveguide gap. Reproduced from Ref. [37] by permission of Optical Society of America

In order to study the effect of the coupler length L on the device performance, we chose three groups of W and G parameters as follows: W = 120 nm and G = 600 nm, W = 140 nm and G = 400 nm, W = 160 nm and G = 600 nm. When L = 40 μm, the minimum TE coupling loss in each W can be obtained by choosing these three groups of parameters. Figures 5.20 and 5.21 show the effect of the device length L on the TE and TM coupling loss, respectively. With the increase of L, the mode conversion loss will decrease. However, because the light limitation in the narrow waveguide is not strong, the surface roughness of the waveguide leads to larger scattering loss and the propagation loss of the waveguide will increase sharply. As a combination result of these two loss, the coupling loss is generally increasing. Here,

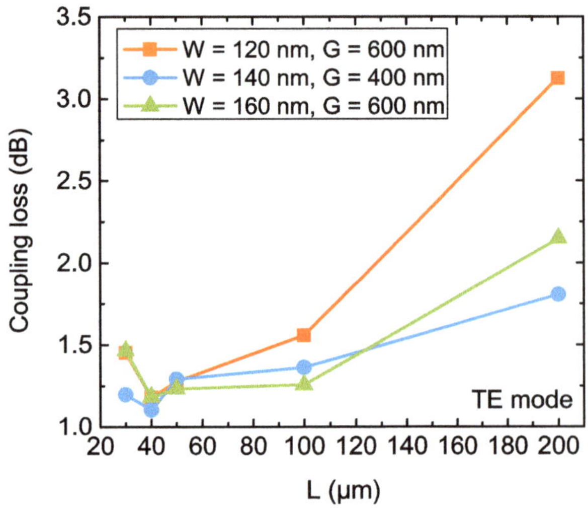

Fig. 5.20 Relationship between the TE coupling loss and coupling length. [2016] IEEE. Reprinted, with permission, from Ref. [38]

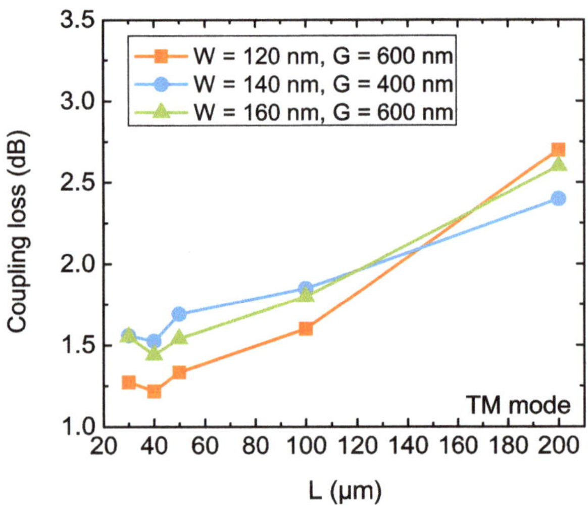

Fig. 5.21 Relationship between the TM coupling loss and coupling length. [2016] IEEE. Reprinted, with permission, from Ref. [38]

it is worth mentioning that for similar device based on narrow waveguide, the effect of waveguide propagation loss on the actual devices must be taken into account.

Considering the minimum TE coupling loss at 1550 nm wavelength, we finally choose the size of inverse taper as follows: W = 140 nm, G = 400 nm and L = 40 μm. The TE and TM coupling loss of the device are only 1.10 dB/facet and 1.52 dB/facet respectively. Figure 5.22 shows the wavelength dependence of the performance of this edge coupler. For the TE mode, the coupling loss is only 0.3 dB in the wavelength range of 1470 to 1630 nm. For the TM mode, the performance of the device will slightly decrease, probably because the optical splitter is mainly optimized in the TE mode. Nevertheless, in this 160 nm wavelength range, there is still only 0.4 dB

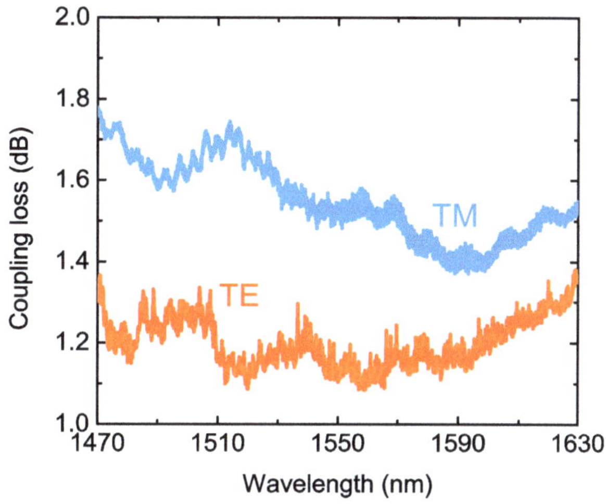

Fig. 5.22 Wavelength dependence of coupling loss when W = 140 nm, G = 400 nm and L = 40 μm. Reproduced from Ref. [37] by permission of Optical Society of America

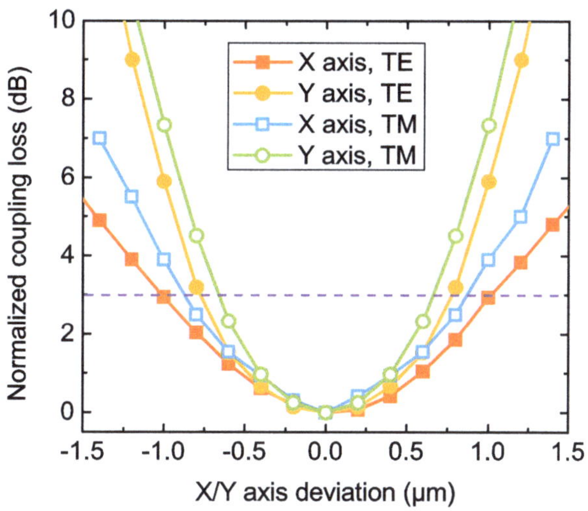

Fig. 5.23 Performance degradation of coupling loss because of the position misalignment of the input fiber. Reproduced from Ref. [37] by permission of Optical Society of America

coupling loss variation. Some small fluctuation in the curve mainly come from the Fabry-Perot reflector or the inherent mechanical jitter of the test platform.

We also studied the effect of fiber misalignment on coupling loss in the horizontal (X) and vertical (Y) directions, which is very important for packaging of optical fiber and chip. Figure 5.23 shows the effect on normalized coupling loss when lensed fiber with MFD = 3.3 μm is used. It is found that the misalignment tolerance of the TE mode is better than that of the TM mode. For the TE and TM modes, the 3 dB loss penalty in the X direction are ±1 μm and ±0.9 μm, respectively. For the TE and TM modes, the 3 dB loss penalty in the Y direction are ±0.75 μm and ±0.9 μm, respectively.

5.5 Conclusion

In this chapter, we study the fiber-to-chip coupling problem in silicon-based optical interconnections. It mainly includes:

1. We have designed and demonstrated a series of fully CMOS-compatible single-tip inverse tapers for edge coupling. We use the SiO_2 gap and deep etching method between the taper tip and the edge of the chip to process the high-quality taper tip, and study the effect of the design parameters on the coupling efficiency. We found that the device length has a obvious effect on the device coupling loss, mainly due to the large waveguide propagation loss. Finally, for a inverse taper with a taper tip width of 0.16 μm and a length of 40 μm, the coupling loss of TE and TM at the wavelength of 1550 nm are 1.13 dB/facet and 1.81 dB/facet respectively. The −1 dB bandwidth is larger than 180 nm.
2. We designed and demonstrated an ultra-compact polarization-insensitive optical combiner. The working principle is based on the two-mode interference. The length of the device is only 5.1 μm, and the excess loss of the two polarizations is both less than 0.1 dB.
3. We designed and demonstrated a novel double-tip inverse taper which consists of two inverse tapers and an ultra-compact polarization-insensitive optical combiner. We have studied the effect of design parameters on coupling loss. And a edge coupler with only 40 μm length is proved. When the wavelength is 1550 nm, the coupling loss for TE and TM are 1.10 dB/facet and 1.52 dB/facet respectively, and the −1 dB bandwidth is more than 160 nm. Its processing is completely CMOS compatible, and it can be further optimized by the previous multi-layer, multi-stage and cantilever technology.

References

1. Loh TH, Wang Q, Zhu J, Ng KT, Lai YC, Huang Y, Ho ST (2010) Ultra-compact multilayer Si/SiO_2 GRIN lens mode-size converter for coupling single-mode fiber to Si-wire waveguide. Opt Express 18:21519–21533
2. Loh TH, Wang Q, Ng KT, Lai YC, Ho ST (2012) CMOS compatible integration of Si/SiO_2 multilayer GRIN lens optical mode size converter to Si wire waveguide. Opt Express 20:14769–14778
3. Almeida VR, Panepucci RR, Lipson M (2003) Nanotaper for compact mode conversion. Opt Lett 28:1302–1304
4. Cardenas J, Poitras CB, Luke K, Luo LW, Morton PA, Lipson M (2014) High coupling efficiency etched facet tapers in silicon waveguides. IEEE Photon Technol Lett 26:2380–2382
5. Shoji T, Tsuchizawa T, Watanabe T, Yamada K, Morita H (2002) Low loss mode size converter from 0.3 μm square Si wire waveguides to singlemode fibres. Electron Lett 38:1669–1670
6. Tsuchizawa T, Yamada K, Fukuda H, Watanabe T, Takahashi JI, Takahashi M, Shoji T, Tamechika E, Itabashi SI, Morita H (2005) Microphotonics devices based on silicon microfabrication technology. Electron Lett 11:232–240
7. McNab SJ, Moll N, Vlasov YA (2003) Ultra-low loss photonic integrated circuit with membrane-type photonic crystal waveguides. Opt Express 11:2927–2939

8. Wahlbrink T, Tsai WS, Waldow M, Forst M, Bolten J, Mollenhauer T, Kurz H (2009) Fabrication of high efficiency SOI taper structures. Microelectron Eng 86:1117–1119
9. Pu M, Liu L, Ou H, Yvind K, Hvam JM (2010) Ultra-low-loss inverted taper coupler for silicon-on-insulator ridge waveguide. Opt Commun 283:3678–3682
10. Ren G, Chen S, Cheng Y, Zhai Y (2011) Study on inverse taper based mode transformer for low loss coupling between silicon wire waveguide and lensed fiber. Opt Commun 284:4782–4788
11. Barkai A, Liu A, Kim D, Cohen R, Elek N, Chang HH, Malik BH, Gabay R, Jones R, Paniccia M, Izhaky N (2008) Double-stage taper for coupling between SOI waveguides and single-mode fiber. J Lightwave Technol 26:3860–3865
12. Na N, Yin T (2012) Misalignment-tolerant spot-size converter for efficient coupling between single-mode fibers and integrated optical receivers. IEEE Photon J 4:187–193
13. Shiraishi K, Yoda H, Ohshima A, Ikedo H, Tsai CS (2007) A silicon-based spot-size converter between single-mode fibers and Si-wire waveguides using cascaded tapers. Appl Phys Lett 91:141120
14. Yoda H, Shiraishi K, Ohshima A, Ishimura T, Furuhashi H, Tsuchiya H, Tsai CS (2009) A two-port single-mode fiber-silicon wire waveguide coupler module using spot-size converters. Appl Phys Lett 27:1315–1319
15. Liao CW, Yang YT, Huang SW, Lee MCM (2011) Fiber-core-matched three-dimensional adiabatic tapered couplers for integrated photonic devices. J Lightwave Technol 29:770–774
16. Khilo A, Popović MA, Araghchini M, Kärtner FX (2010) Efficient planar fiber-to-chip coupler based on two-stage adiabatic evolution. J Lightwave Technol 18:15790–15806
17. Park H, Kim S, Park J, Joo J, Kim G (2013) A fiber-to-chip coupler based on Si/SiON cascaded tapers for Si photonic chips. Opt Express 21:29313–29319
18. Tokushima M, Kamei A, Horikawa T (2012) Dual-tapered 10 μm spot-size converter with double core for coupling polarization-independent silicon rib waveguides to single-mode optical fibers. Appl Phys Express 5:022202
19. Shiraishi K, Yoda H, Tsai CS (2012) A two-port polarization-insensitive coupler module between single-mode fiber and silicon-wire waveguide. Opt Express 20:24370–24375
20. Ku KN, Lee MCM (2013) Wide-band optical mode converters for coupling between fibers and silicon photonic wires with large misalignment tolerance. J Lightwave Technol 31:1616–1620
21. Sun P, Reano RM (2009) Cantilever couplers for intra-chip coupling to silicon photonic integrated circuits. Opt Express 17:4565–4574
22. Wood M, Sun P, Reano RM (2012) Compact cantilever couplers for low-loss fiber coupling to silicon photonic integrated circuits. Opt Express 20:164–172
23. Bakir BB, De Gyves AV, Orobtchouk R, Lyan P, Porzier C, Roman A, Fedeli JM (2010) Low-loss (<1 dB) and polarization-insensitive edge fiber couplers fabricated on 200-mm silicon-on-insulator wafers. IEEE Photon Technol Lett 22:739–741
24. Chen L, Doerr CR, Chen YK, Liow TY (2010) Low-loss and broadband cantilever couplers between standard cleaved fibers and high-index-contrast Si_3N_4 or Si waveguides. IEEE Photon Technol Lett 22:1744–1746
25. Fang Q, Liow TY, Song JF, Tan CW, Yu MB, Lo GQ, Kwong DL (2010) Suspended optical fiber-to-waveguide mode size converter for silicon photonics. Opt Express 18:7763–7769
26. Fang Q, Song J, Luo X, Yu M, Lo G, Liu Y (2011) Mode-size converter with high coupling efficiency and broad bandwidth. Opt Express 19:21588–21594
27. Jia L, Song J, Liow TY, Luo X, Tu X, Fang Q, Koh SC, Yu M, Lo G (2014) Mode size converter between high-index-contrast waveguide and cleaved single mode fiber using SiON as intermediate material. Opt Express 22:23652–23660
28. Tao SH, Song J, Fang Q, Yu MB, Lo GQ, Kwong DL (2008) Improving coupling efficiency of fiber-waveguide coupling with a double-tip coupler. Opt Express 16:20803–20808
29. Tu X, Fu H, Geng D (2014) Y-branch edge coupler between cleaved single mode fiber and nano-scale waveguide on silicon-on-insulator platform. In: Asia communications and photonics conference. Optical Society of America
30. Liu Y, Yu J (2007) Low-loss coupler between fiber and waveguide based on silicon-on-insulator slot waveguides. Appl Opt 46:7858–7861

31. Galan JV, Sanchis P, Blasco J, Martinez A, Marti J (2008) High efficiency fiber coupling to silicon sandwiched slot waveguides. Opt Commun 281:5173–5176
32. Cheben P, Xu DX, Janz S, Densmore A (2006) Subwavelength waveguide grating for mode conversion and light coupling in integrated optics. Opt Express 14:4695–4702
33. Min TENG, Niu B, Han K, Qi M (2015) Effect of waveguide surface roughness on the fiber coupling efficiency of inverse tapers. In: Optical fiber communication conference. Optical Society of America
34. Sheng Z, Wang Z, Qiu C, Li L, Pang A, Wu A, Wang X, Zou S, Gan F (2012) A compact and low-loss MMI coupler fabricated with CMOS technology. IEEE Photon J 4:2272–2277
35. Zhang Y, Yang S, Lim AEJ, Lo GQ, Galland C, Baehr-Jones T, Hochberg M (2013) A compact and low loss Y-junction for submicron silicon waveguide. Opt Express 21:1310–1316
36. Xiao Z, Luo X, Lim PH, Prabhathan P, Silalahi ST, Liow TY, Zhang J, Luan F (2013) Ultra-compact low loss polarization insensitive silicon waveguide splitter. Opt Express 21:16331–16336
37. Wang J, Xuan Y, Lee C, Niu B, Liu L, Liu GN, Qi M (2016) Low-loss and misalignment-tolerant fiber-to-chip edge coupler based on double-tip inverse tapers. In: Optical fiber communication conference. Optical Society of America
38. Wang J, Xuan Y, Qi M, Liu L, Liu GN (2016) Comparison and analysis on single-layer Si fiber-to-chip edge couplers with different taper tips. In: IEEE 13th international conference on group IV photonics (GFP)

Chapter 6
Summary and Future Work

This dissertation mainly focuses on four important scientific problems in silicon photonics, i.e., silicon electro-optic modulator, advanced multiplexing mechanism, polarization controlling and fiber-to-chip coupling,

1. In the research of silicon electro-optic modulator devices, the design theory of silicon electro-optic modulator was put forward systematically. The method of optimizing the waveguide size in the doping region was introduced through the combination of semiconductor process simulation and in-house programming. The effect of the position of PN junction on the modulation efficiency and absorption loss was found. Then the full analytical model of the equivalent circuit of the modulator with the lateral PN junction and the extraction of each circuit element were studied in detail. The size of the traveling-wave electrode was optimized by using the theory of microwave transmission line. Finally, we used the commercial 130 nm CMOS process to demonstrate a high-speed carrier-depletion silicon electro-optic modulator based on the MZI structure, and carried out a comprehensive performance characterization of the samples with different lengths. The measurement included the transmission spectrum characterization, the S-parameter test of the traveling-wave electrode, the electro-optic frequency response test, the high-speed optical eye test and the bi-error-ratio test. The relevant research results were published in Journal of Lightwave Technology and I made a oral report on the Asia Communications and Photonics (ACP) Conference in November 2012. In addition, I implemented a low-power 30 Gb/s carrier-depletion silicon electro-optical modulator based on MRR. The footprint of the device is only $30 \times 20\ \mu m^2$, and the power consumption is only 167 fJ/bit, which can be further reduced.

J. Wang, *CMOS-Compatible Key Engineering Devices for High-Speed Silicon-Based Optical Interconnections*, Springer Theses,
https://doi.org/10.1007/978-981-13-3378-1_6

2. In regard to the advanced multiplexing mechanism of silicon-based optical interconnections, we have investigated the following three points.

   a. In the research of WDM devices, we have successfully demonstrated high-performance saddle-shaped Si-AWG devices fabricated with commercial 130 nm CMOS process. The channel spacing included 0.8, 1.6, 3.2 and 8 nm. We took a 3.2 nm Si-AWG as an example. Through comparative study, we found that a comprehensive optimization method of broadening the arrayed waveguide and introducing a shallow-etching mode converter near FPR can improve the device performance more effectively than using a single optimization method. Based on this, we successfully implemented a Si-AWG router with channel spacing of 0.8 nm for DWDM applications. Rotating cycle test results show that the two devices worked well. We further tested the transmission characteristics of two AWGs in a 3-channel WDM system, including the bit-error rate and the optical eye diagram of the demultiplexed signal. The relevant research results were published in Optics Express. One reviewer comments, *The authors present an interesting study on silicon photonic AWG routers with 3.2 and 0.8 nm channel spacing by optimizing the design of the AWG...The whole manu. is well written, and I recommend publishing it on OE with no further revision required...* Finally, we also demonstrated Si-AWG devices with other channel spacing and presented the preliminary results of flat-top Si-AWGs.
   b. In the research of PDM devices, we first proposed a scheme to achieve ultra-broadband PBS by using the mode-sorting effect of asymmetric Y-junction. We achieved different mode-sorting functions by cascading three Y-junctions and carefully optimizing their design parameters. Finally, the overall device exhibits excellent performance in the wavelength range from 1450 to 1750 nm. We further discussed the effect of some common process errors on device performances, and put forward some feasible ideas for further performance optimization. The relevant research results were published in IEEE Photonics Journal. One reviewer comments, *The authors present an interesting new design of polarization beam splitter based on cascade asymmetric Y-junction divider on SOI substrate.*
   c. In the research of the MDM devices, we used the counter-tapered coupler to realize a high-performance MDM link, which has the advantages of broadband properties, high process tolerance and extensibility. We analyzed and optimized the MDM multiplexer based on counter-tapered coupler, characterized a two-mode MDM link and its extended three-mode MDM link. For the two-mode MDM link, the -1 dB bandwidth is larger than 180 nm, the crosstalk is less than -13 dB in the wavelength range from 1460 to 1640 nm, and it is stable for the process error (width variation -60 to 40 nm) and the temperature change ($-25\,^{\circ}$C to $75\,^{\circ}$C). For the three-mode MDM link, the crosstalk is less than -10 dB at 180 nm wavelength range. The relevant

research results were published in Optics Letters. One reviewer comments, *...The manuscript is logically organized and reasonably well written, and describes results of interest in the field, both from simulation and with experimental corroboration.*

3. In the research aspect of polarization-controlling device, we have studied the polarization splitter-rotator in the following aspects.

   a. We proposed a ultra-compact PSR device based on rib directional coupler with SiO_2cladding, which can relieve the limitation of the traditional PSR that can not be processed with the CMOS process. The coupling length is only 9 μm, and the total length is only 25 μm. The relevant research results were published in Optics Express. One reviewer comments, *The paper clearly presented a new design of compact polarization splitter-rotator (PSR) based on asymmetric rib directional coupler on SOI with an oxide top cladding...the proposed structure still shows significant improvement...*
   b. We used cascaded or assisted mode converters to improve the PSR devices based on rib directional coupler. Simulation results show that the device performance and process tolerance have been improved. Based on the design of the assisted mode converter, we used the bi-level tapered mode converter instead of the rib mode converter to process the PSR device based on DC, and built the test set-up for the polarization controlling devices.
   c. The design of ulra-broadband PSR devices was proposed by using the three-order linear mode converter and the mode-sorting asymmetric Y-junction. The simulation results show that the device's working range is as high as 400 nm, which is sufficient to cover all optical communication bands. The relevant research results were published in Optics Express. One reviewer comments, *I find this a well written paper, presenting a novel design. The concept is straightforward and easy to understand, but I have not seen exactly this design presented before...The manuscript represents a solid piece of work and the results presented in this paper are interesting and significant. Since this topic is of great interest to photonic circuits communities and the presented results are of high technical quality and relevance...*
   d. We used cascaded MMI couplers, bi-level mode converter and a 90-degree phase shifter to design a PSR device with high fabrication tolerance. This device has a high tolerance on various process errors, and can meet the requirements of CMOS process. It also has good performance and can adapt to the requirements of working in a certain band. The relevant research results were published in Optics Express. One reviewer comments, *Overall, the manuscript is presented well and the achieved results are interesting...the authors proposed a* novel *silicon-on-insulator (SOI) polarization splitter-rotator (PSR)...*
   e. We extended the working range of PSR devices from the near infrared optical communication band to the mid-infrared range. Considering the role of SiN in reducing the absorption loss, a series of PSR devices for different applications were proposed. The relevant research results were published

in Optics Express. One reviewer comments, *The paper is well written, and the work presented is sound...It will contribute to the body of work on integrated photonics for the mid-IR...the reviewer believes the investigation of these designs in the mid-infrared range is novel and useful to the integrated optics community.*

4. In terms of fiber-to-chip coupling, we have studied the following three aspects.
   a. We designed and implemented a series of fully CMOS-compatible single-tip inverse tapers. We used the SiO_2 gap and deep etching method between the taper tip and the edge of the chip to achieve high-quality taper tip, and studied the effect of the design parameters on the coupling efficiency. We found that the device length has considerable effect on the coupling loss because of the large waveguide propagation loss. For a inverse coupler with a taper tip width of 0.16 μm and a length of 40 μm, the coupling loss of TE and TM at 1550 nm wavelength is 1.13 dB/facet and 1.81 dB/facet respectively, and the -1 dB bandwidth is larger than 180 nm.
   b. We designed and implemented an ultra-compact polarization-insensitive optical splitter based on two-mode interference. The length of the device is only 5.1 μm, and the excess loss of the two polarization are both less than 0.1 dB.
   c. A novel double-tip inverse taper was designed and implemented, which consists of two normal single-tip inverse tapers and a ultra-compact polarization-insensitive optical combiner. We have studied the effect of design parameters on coupling loss. And a coupler with only 40 μm length was proved. When the wavelength is 1550 nm, the TE and TM coupling loss is 1.10 dB/facet and 1.52 dB/facet respectively. The -1 dB bandwidth is more than 160 nm.

The starting point of all the work in this dissertation is CMOS-compatibility, and the involved devices are necessary for silicon photonic communication chips. Therefore, it is hoped that the work in this dissertation can help the researchers and engineers engaged in silicon photonics.

There will be still many works to be done in the research and development of silicon photonics.

1. In the study of the modulator, the design optimization of the novel traveling-wave electrode and the analysis of signal integrity are still absent. It can be further studied referring to the related experience of the lithium niobate modulator. At the same time, an engineering device design process needs to be established to standardize the design and test of the modulator.
2. In the advanced multiplexing schemes, large-scale hybrid multiplexing may be a hot topic in the future. With the stability of CMOS process, it is possible to achieve high-speed transmission of hundreds of channels of Tb/s. At the same time, it is necessary to establish a corresponding high-speed test platform to evaluate the performances in the communication system.

3. In the aspect of polarization splitter-rotator, although we have proposed many devices, some device samples have not yet been tested, which need to be completed in the future.
4. In terms of the fiber-to-chip coupling, it is necessary to further explore low-cost industrial packaging methods.
5. By the end of May 2015, the research and development of silicon photonics in China still remained at the verification stage of single component. Therefore, there is still a long way to go in achieving silicon-based optoelectronic integration.
6. There has been no significant progress in the research of pure silicon lasers, so it is necessary to enhance the study of hybrid lasers and other silicon-based CMOS-compatible lasers. Otherwise, it will be a weakness for large-scale applications of silicon photonics.
7. The devices in silicon-based optical interconnections have been basically finalized and the performance is steadily improved, but there are many interesting topics to be studied in optical routing, optical switching and optical logic based on silicon photonic technology. Especially large-scale reconfigurable all-optical switching array may be one of the key technologies for the next-generation optical communication. In addition, there have been a lot of work to combine silicon photonics with photonic crystals, but there are few reports on physical phenomena in graphene, artificial medium materials and plasmonics. Nevertheless, it can be predicted that this will be a good basic research direction. Moreover, the advantage of silicon photonics technology in large-scale signal processing is still expanding in the applications of mid-infrared sensing, microwave photonics and quantum communication.

MIX
Papier aus verantwortungsvollen Quellen
Paper from responsible sources
FSC® C105338

If you have any concerns about our products,
you can contact us on
ProductSafety@springernature.com

In case Publisher is established outside the EU,
the EU authorized representative is:
Springer Nature Customer Service Center GmbH
Europaplatz 3, 69115 Heidelberg, Germany

Printed by Libri Plureos GmbH
in Hamburg, Germany